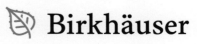

**Advanced Courses in Mathematics
CRM Barcelona**

Centre de Recerca Matemàtica

Managing Editor:
Enric Ventura

More information about this series at http://www.springer.com/series/5038

Vlad Bally • Lucia Caramellino • Rama Cont

Stochastic Integration by Parts and Functional Itô Calculus

Editors for this volume:
Frederic Utzet, Universitat Autònoma de Barcelona
Josep Vives, Universitat de Barcelona

Vlad Bally
Université de Marne-la-Vallée
Marne-la-Vallée, France

Rama Cont
Department of Mathematics
Imperial College
London, UK

and

Centre National de Recherche Scientifique (CNRS)
Paris, France

Lucia Caramellino
Dipartimento di Matematica
Università di Roma "Tor Vergata"
Roma, Italy

Rama Cont dedicates his contribution to Atossa, for her kindness and constant encouragement.

ISSN 2297-0304   ISSN 2297-0312   (electronic)
Advanced Courses in Mathematics - CRM Barcelona
ISBN 978-3-319-27127-9   ISBN 978-3-319-27128-6   (eBook)
DOI 10.1007/978-3-319-27128-6

Library of Congress Control Number: 2016933913

Printed on acid-free paper

This book is published under the trade name Birkhäuser.
The registered company is Springer International Publishing AG (www.birkhauser-science.com)

Foreword

During July 23th to 27th, 2012, the first session of the *Barcelona Summer School on Stochastic Analysis* was organized at the Centre de Recerca Matemàtica (CRM) in Bellaterra, Barcelona (Spain). This volume contains the lecture notes of the two courses given at the school by Vlad Bally and Rama Cont.

The notes of the course by Vlad Bally are co-authored with her collaborator Lucia Caramellino. They develop integration by parts formulas in an abstract setting, extending Malliavin's work on abstract Wiener spaces, and thereby being applicable to prove absolute continuity for a broad class of random vectors. Properties like regularity of the density, estimates of the tails, and approximation of densities in the total variation norm are considered. The last part of the notes is devoted to introducing a method to prove existence of density based on interpolation spaces. Examples either not covered by Malliavin's approach or requiring less regularity are in the scope of its applications.

Rama Cont's notes are on Functional Itô Calculus. This is a non-anticipative functional calculus extending the classical Itô calculus to path-dependent functionals of stochastic processes. In contrast to Malliavin Calculus, which leads to *anticipative* representation of functionals, with Functional Itô Calculus one obtains *non-anticipative* representations, which may be more natural in many applied problems. That calculus is first introduced using a pathwise approach (that is, without probabilities) based on a notion of directional derivative. Later, after the introduction of a probability on the space of paths, a weak functional calculus emerges that can be applied without regularity conditions on the functionals. Two applications are studied in depth; the representation of martingales formulas, and then a new class of path-dependent partial differential equations termed *functional Kolmogorov equations*.

We are deeply indebted to the authors for their valuable contributions. Warm thanks are due to the Centre de Recerca Matemàtica, for its invaluable support in the organization of the School, and to our colleagues, members of the Organizing Committee, Xavier Bardina and Marta Sanz-Solé. We extend our thanks to the following institutions: AGAUR (Generalitat de Catalunya) and Ministerio de Economía y Competitividad, for the financial support provided with the grants SGR 2009-01360, MTM 2009-08869 and MTM 2009-07203.

Frederic Utzet and Josep Vives

Contents

II   Functional Itô Calculus and Functional Kolmogorov Equations
Rama Cont   **115**

Part I

Integration by Parts Formulas, Malliavin Calculus, and Regularity of Probability Laws

Vlad Bally and Lucia Caramellino

Preface

The lectures we present here turn around the following integration by parts formula. On a probability space $(\Omega, \mathcal{F}, \mathbb{P})$ we consider a random variable G, and a random vector $F = (F_1, \ldots, F_d)$ taking values in $\mathbb{R}^d$. Moreover, we consider a multiindex $\alpha = (\alpha_1, \ldots, \alpha_m) \in \{1, \ldots, d\}^m$ and we write $\partial_\alpha = \partial_{x^{\alpha_1}} \cdots \partial_{x^{\alpha_m}}$. We look for a random variable, which we denote by $H_\alpha(F; G) \in L^p(\Omega)$, such that the following integration by parts formula holds:

$$\mathrm{IBP}_{\alpha,p}(F, G) \qquad \mathbb{E}(\partial_\alpha f(F)G) = \mathbb{E}(f(F)H_\alpha(F; G)), \quad \forall f \in C_b^\infty(\mathbb{R}^d),$$

where $C_b^\infty(\mathbb{R}^d)$ denotes the set of infinitely differentiable functions $f : \mathbb{R}^d \to \mathbb{R}$ having bounded derivatives of any order. This is a set of test functions, that can be replaced by other test functions, such as the set $C_c^\infty(\mathbb{R}^d)$ of infinitely differentiable functions $f : \mathbb{R}^d \to \mathbb{R}$ with compact support.

The interest in this formula comes from the Malliavin calculus; this is an infinite-dimensional differential calculus associated to functionals on the Wiener space which permits to construct the weights $H_\alpha(F; G)$ in the above integration by parts formula. We will treat several problems related to $\mathrm{IBP}_{\alpha,p}(F, G)$ that we list now.

Problem 1

Suppose one is able to produce $\mathrm{IBP}_{\alpha,p}(F, G)$, does not matter how. What can we derive from this? In the classical Malliavin calculus such formulas with $G = 1$ have been used in order to:

(1) prove that the law of F is absolutely continuous and, moreover, to study the regularity of its density;

(2) give integral representation formulas for the density and its derivatives;

(3) obtain estimates for the tails of the density, as well;

(4) obtain similar results for the conditional expectation of G with respect to F, assuming If $\mathrm{IBP}_{\alpha,p}(F, G)$ holds (with a general G).

Our first aim is to derive such properties in the abstract framework that we describe now.

A first remark is that $\mathrm{IBP}_{\alpha,p}(F, G)$ does not involve random variables, but the law of the random variables: taking conditional expectations in the above formula we get $\mathbb{E}(\partial_\alpha f(F)\mathbb{E}(G \mid \sigma(F))) = \mathbb{E}(f(F)\mathbb{E}(H_\alpha(F; G) \mid \sigma(F)))$. So, if we denote $g(x) = \mathbb{E}(G \mid F = x)$, $\theta_\alpha(g)(x) = \mathbb{E}(H_\alpha(F; G) \mid F = x)$, and if $\mu_F(dx)$ is the law of F, then the above formula reads

$$\int \partial_\alpha f(x) g(x) d\mu_F(x) = \int f(x) \theta_\alpha(g)(x) d\mu_F(x).$$

If $\mu_F(dx)$ is replaced by the Lebesgue measure, then this is a standard integration by parts formula and the theory of Sobolev spaces comes on. But here, we have the specific point that the reference measure is μ_F, the law of F. We denote $\partial_\alpha^{\mu_F} g = \theta_\alpha(g)$ and this represents somehow a weak derivative of g. But it does not verify the chain rule, so it is not a real derivative. Nevertheless, we may develop a formalism which is close to that of the Sobolev spaces, and express our results in this formalism. Shigekawa [45] has already introduced a very similar formalism in his book, and Malliavin introduced the concept of *covering vector fields*, which is somehow analogous. The idea of giving an abstract framework related to integration by parts formulas already appears in the book of Bichteler, Gravereaux and Jacod [16] concerning the Malliavin calculus for jump processes.

A second ingredient in our approach is to use the Riesz representation formula; this idea comes from the book of Malliavin and Thalmaier [36]. So, if Q_d is the Poisson kernel on $\mathbb{R}^d$, i.e., the solution of $\triangle Q_d = \delta_0$ (δ_0 denoting the Dirac mass), then a formal computation using $\mathrm{IBP}_{\alpha,p}(F, 1)$ gives

$$p_F(x) = \mathbb{E}(\delta_0(F - x)) = \sum_{i=1}^d \mathbb{E}\big(\partial_i^2 Q_d(F - x)\big) = \sum_{i=1}^d \mathbb{E}\big(\partial_i Q_d(F - x) H_i(F, 1)\big).$$

This is the so-called Malliavin–Thalmaier representation formula for the density. One may wonder why do we perform one integration by parts and not two? The answer is that Q_d is singular in zero and then $\partial_i^2 Q_d$ is "not integrable" while $\partial_i Q_d$ is "integrable"; so, as we expect, integration by parts permits to regularize things. As suggested before, the key issue when using Q_d is to handle the integrability problems, and this is an important point in our approach. The use of the above representation allows us to ask less regularity for the random variables at hand (this has already been remarked and proved by Shigekawa [45] and by Malliavin [33]).

All these problems —(1), (2), (3), and (4) mentioned above— are discussed in Chapter 1. Special attention is given to *localized integration by parts formulas* which, roughly speaking, means that we take G to be a smooth version of $1_{\{F \in A\}}$. This turns out to be extremely useful in a number of problems.

Problem 2

How to construct $\mathrm{IBP}_{\alpha,p}(F, G)$? The strategy in Malliavin calculus is roughly speaking as follows. One defines a differential operator D on a subspace of "regular objects" which is dense in the Wiener space, and then takes the standard extension of this unbounded operator. There are two main approaches: the first one is to consider the expansion in Wiener chaos of the functionals on the Wiener space. In this case the multiple stochastic integrals are the "regular objects" on which D is defined directly. We do not take this point of view here (see Nualart [41] for a complete theory in this sense). The second one is to consider the cylindrical functions of the Brownian motion W as "regular objects", and this is our point of view. We consider $F_n = f(\Delta_n^1, \ldots, \Delta_n^{2^n})$ with $\Delta_n^k = W(k/2^n) - W((k-1)/2^n)$ and f being a smooth function, and we define $D_s F = \partial_k f(\Delta_n^1, \ldots, \Delta_n^{2^n})$ if $(k-1)/2^n \le s < k/2^n$. Then, given a general functional F, we look for a sequence F_n of simple functionals such that $F_n \to F$ in $L^2(\Omega)$, and if $DF_n \to U$ in $L^2([0,1] \times \Omega)$, then we define $D_s F = U_s$.

Looking to the "duality formula" for DF_n (which is the central point in Malliavin calculus), one can see that the important point is that we know explicitly the density p of the law of $(\Delta_n^1, \ldots, \Delta_n^{2^n})$ (Gaussian), and this density comes on in the calculus by means of $\ln p$ and of its derivatives only. So we may mimic all the story for a general finite-dimensional random vector $V = (V_1, \ldots, V_m)$ instead of $\Delta = (\Delta_n^1, \ldots, \Delta_n^{2^n})$. This is true in the finite-dimensional case (for simple functionals), but does not go further to general functionals (the infinite-dimensional calculus). Nevertheless, we will see in a moment that, even without passing to the limit, integration by parts formulas are useful. This technology has already been used in the framework of jump type diffusions in [8, 10]. In Chapter 2, and specifically in Section 2.1, we present this "finite-dimensional abstract Malliavin calculus" and then we derive the standard infinite-dimensional Malliavin calculus. Moreover, we use these integration by parts formulas and the general results from Chapter 1 in order to study the regularity of the law in this concrete situation. But not only this: we also obtain quantitative estimates concerning the density of the law as mentioned in the points (1), (2), (3), and (4) from Problem 1.

Problem 3

There are many other applications of the integration by parts formulas in addition to the regularity of the law. We mention here few ones.

Malliavin calculus permits to obtain results concerning the rate of convergence of approximation schemes (for example the Euler approximation scheme for diffusion processes) for test functions which are not regular but only measurable and bounded. And moreover, under more restrictive hypotheses and with a little bit more effort, to prove the convergence of the density functions; see for example [12, 13, 27, 28, 30]. Another direction is the study of the density in short time (see, e.g., Arous and Leandre [14]), or the strict positivity of the density (see, e.g., Bally and Caramellino [3]). We do not treat here these problems, but we give a result which is central in such problems: in Section 2.4 we provide an estimate of the distance between the density functions of two random variables in terms of the weights $H_\alpha(F, G)$, which appear in the integration by parts formulas. Moreover we use this estimates in order to give sufficient conditions allowing one to obtain convergence in the total variation distance for a sequence of random variables which converge in distribution. The localization techniques developed in Chapter 1 play a key role here.

Problem 4

We present an alternative approach for the study of the regularity of the law of F. The starting point is the paper by Fournier and Printems [26]. Coming back to the approach presented above, we recall that we have defined $D_s F = \lim_n D_s F_n$ and then we used DF in order to construct $\text{IBP}_{\alpha,p}(F,1)$. But one may proceed in an alternative way: since DF_n is easy to define (it is just a finite-dimensional gradient), one may use elementary integration by parts formulas (in finite-dimensional spaces) in order to obtain $\text{IBP}_{\alpha,p}(F_n,1)$. Then passing to the limit $n \to \infty$ in $\text{IBP}_{\alpha,p}(F_n,1)$, one obtains $\text{IBP}_{\alpha,p}(F,1)$. If everything works well, then we are done —but we are not very far from the Malliavin calculus itself. The interesting fact is that sometimes this argument still works even if things are "going bad", that is, even when $H_\alpha(F_n,1) \uparrow \infty$. The idea is the following. We denote by $\widehat{p}_F$ the Fourier transform of F and by $\widehat{p}_{F_n}$ the Fourier transform of F_n. If we are able to prove that $\int |\widehat{p}_F(\xi)|^2 \, d\xi < \infty$, then the law of F is absolutely continuous. In order to do it we notice that $\partial_x^m e^{i\xi x} = (i\xi)^m e^{i\xi x}$ and we use $\text{IBP}_{m,p}(F_n,1)$ in order to obtain

$$\widehat{p}_{F_n}(\xi) = \frac{1}{(i\xi)^m} \mathbb{E}\big(\partial_x^m e^{i\xi F_n}\big) = \frac{1}{(i\xi)^m} \mathbb{E}\big(e^{i\xi F_n} H_m(F_n,1)\big).$$

Then, for each $m \in \mathbb{N}$,

$$|\widehat{p}_F(\xi)| \le |\widehat{p}_F(\xi) - \widehat{p}_{F_n}(\xi)| + |\widehat{p}_{F_n}(\xi)| \le |\xi| \, \mathbb{E}\big(|F - F_n|\big) + \frac{1}{|\xi|^m} \mathbb{E}\big(|H_m(F_n,1)|\big).$$

So, if we obtain a good balance between $\mathbb{E}(|F - F_n|) \downarrow 0$ and $\mathbb{E}(|H_m(F_n,1))| \uparrow \infty$, we have a chance to prove that $\int_{\mathbb{R}^d} |\widehat{p}_F(\xi)|^2 \, d\xi < \infty$ and so to solve our problem. One unpleasant point in this approach is that it depends strongly on the dimension d of the space: the above integral is convergent if $|\widehat{p}_F(\xi)| \le \text{Const} \, |\xi|^{-\alpha}$ with $\alpha > d/2$, and this may be rather restrictive for large d. In [20], Debussche and Romito presented an alternative approach which is based on certain relative compactness criterion in Besov spaces (instead of the Fourier transform criterion), and their method is much more performing. Here, in Chapter 3 we give a third approach based on an expansion in Hermite series. We also observe that our approach fits in the theory of interpolation spaces and this seems to be the natural framework in which the equilibrium between $\mathbb{E}(|F - F_n|) \downarrow 0$ and $\mathbb{E}(|H_m(F_n,1)|) \uparrow \infty$ has to be discussed.

The class of methods presented above goes in a completely different direction than the Malliavin calculus. One of the reasons is that Malliavin calculus is somehow a pathwise calculus —the approximations $F_n \to F$ and $DF_n \to DF$ are in some L^p spaces and so, passing to a subsequence, one can also achieve almost sure convergence. This allows one to use it as an efficient instrument for analysis on the Wiener space (the central example is the Clark–Ocone representation formula). In particular, one has to be sure that nothing blows up. In contrast, the argument presented above concerns the laws of the random variables at hand and, as mentioned, it allows to handle the blow-up. In this sense it is more flexible and permits to deal with problems which are out of reach for Malliavin calculus. On the other hand, it is clear that if Malliavin calculus may be used (and so real integration by parts formulas are obtained), then the results are more precise and deeper: because one does not only obtain the regularity of the law, but also integral representation formulas, tail estimates, lower bounds for densities and so on. All this seems out of reach with the asymptotic methods presented above.

Conclusion

The results presented in these notes may be seen as a complement to the classical theory of Sobolev spaces on $\mathbb{R}^d$ that are suited to treat problems of regularity of probability laws. We stress that this is different from Watanabe's distribution theory on the Wiener space. Let $F \colon \mathcal{W} \to \mathbb{R}^d$. The distribution theory of Watanabe deals with the infinite-dimensional space $\mathcal{W}$, while the results in these notes concern μ_F, the law of F, which lives in $\mathbb{R}^d$. The Malliavin calculus comes on in the construction of the weights $H_\alpha(F, G)$ in the integration by parts formula (1) and its estimates. We also stress that the point of view here is somehow different from the one in the classical theory of Sobolev spaces: there the underling measure is the Lebesgue measure on $\mathbb{R}^d$, whereas here, if F is given, then the basic measure in the integration by parts formula is μ_F. This point of view comes from Malliavin calculus (even if we are in a finite-dimensional framework). Along these lectures, almost all the time, we fix F. In contrast, the specific achievement of Malliavin calculus (and of Watanabe's distribution theory) is to provide a theory in which one can deal with all the "regular" functionals F in the same framework.

Vlad Bally, Lucia Caramellino

Chapter 1

Integration by parts formulas and the Riesz transform

The aim of this chapter is to develop a general theory allowing to study the existence and regularity of the density of a probability law starting from integration by parts type formulas (leading to general Sobolev spaces) and the Riesz transform, as done in [2]. The starting point is given by the results for densities and conditional expectations based on the Riesz transform given by Malliavin and Thalmaier [36]. Let us start by speaking in terms of random variables.

Let F and G denote random variables taking values on $\mathbb{R}^d$ and $\mathbb{R}$, respectively, and consider the following integration by parts formula: there exist some integrable random variables $H_i(F, G)$ such that for every test function $f \in C_c^\infty(\mathbb{R}^d)$

$$\text{IBP}_i(F, G) \quad \mathbb{E}\big(\partial_i f(F)G\big) = -\mathbb{E}\big(f(F)H_i(F, G)\big), \quad i = 1, \ldots, d.$$

Malliavin and Thalmaier proved that if $\text{IBP}_i(F, 1)$, $i = 1, \ldots, d$, hold and the law of F has a continuous density p_F, then

$$p_F(x) = -\sum_{i=1}^{d} \mathbb{E}(\partial_i Q_d(F - x)H_i(F, 1)),$$

where Q_d denotes the Poisson kernel on $\mathbb{R}^d$, that is, the fundamental solution of the Laplace operator. Moreover, they also proved that if $\text{IBP}_i(F, G)$, $i = 1, \ldots, d$, a similar representation formula holds also for the conditional expectation of G with respect to F. The interest of Malliavin and Thalmaier in these representations came from numerical reasons —they allow one to simplify the computation of densities and conditional expectations using a Monte Carlo method. This is crucial in order to implement numerical algorithms for solving nonlinear PDE's or optimal stopping problems. But there is a difficulty one runs into: the variance of the estimators produced by such a representation formula is infinite. Roughly speaking, this comes from the blowing up of the Poisson kernel around zero: $\partial_i Q_d \in L^p$ for $p < d/(d-1)$, so that $\partial_i Q_d \notin L^2$ for every $d \geq 2$. Hence, estimates of $\mathbb{E}(|\partial_i Q_d(F - x)|^p)$ are crucial in this framework and this is the central point of interest here. In [31, 32], Kohatsu-Higa and Yasuda proposed a solution to this problem using some cut-off arguments. And, in order to find the optimal cut-off level, they used the estimates of $\mathbb{E}(|\partial_i Q_d(F - x)|^p)$ which are proven in Theorem 1.4.1.

So, our central result concerns estimates of $\mathbb{E}(|\partial_i Q_d(F-x)|^p)$. It turns out that, in addition to the interest in numerical problems, such estimates represent a suitable tool for obtaining regularity of the density of functionals on the Wiener space, for which Malliavin calculus produces integration by parts formulas. Before going further, let us mention that one may also consider integration by parts formulas of higher order, that is

$$\mathrm{IBP}_\alpha(F,G) \qquad\qquad \mathbb{E}(\partial_\alpha f(F)) = (-1)^{|\alpha|}\mathbb{E}\big(f(F)H_\alpha(F,G)\big),$$

where $\alpha = (\alpha_1, \ldots, \alpha_k) \in \{1, \ldots, d\}^k$ and $|\alpha| = k$. We say that an integration by parts formula of order k holds if this is true for every $\alpha \in \{1, \ldots, d\}^k$. Now, a first question is: which is the order k of integration by parts that one needs in order to prove that the law of F has a continuous density p_F? If one employs a Fourier transform argument (see Nualart [41]) or the representation of the density by means of the Dirac function (see Bally [1]), then one needs d integration by parts if $F \in \mathbb{R}^d$. In [34] Malliavin proves that integration by parts of order one suffices to obtain a continuous density, independently of the dimension d (he employs some harmonic analysis arguments). A second problem concerns estimates of the density p_F (and of its derivatives) and such estimates involve the L^p norms of the weights $H_\alpha(F,1)$. In the approach using the Fourier transform or the Dirac function, the norms $\|H_\alpha(F,1)\|_p$ for $|\alpha| \leq d$ are involved if one estimates $\|p_F\|_\infty$. However, Shigekawa [45] obtains estimates of $\|p_F\|_\infty$ depending only on $\|H_i(F,1)\|_p$, so on the weights of order one (and similarly for derivatives). To do this, he needs some Sobolev inequalities which he proves using a representation formula based on the Green function, and some estimates of modified Bessel functions. Our program is similar, but the main tools used here are the Riesz transform and the estimates of the Poisson kernel mentioned above.

Let us be more precise. First, we replace μ_F with a general probability measure μ. Then, $\mathrm{IBP}_i(F,G)$ may also be written as

$$\int \partial_i f(x) g(x) \mu(dx) = -\int f(x)\partial_i^\mu g(x)\mu(dx),$$

where $\mu = \mu_F$ (the law of F), $g(x) := \mathbb{E}(G \mid F = x)$ and $\partial_i^\mu g(x) := \mathbb{E}(H(F,G) \mid F = x)$. This suggests that we can work in the abstract framework of Sobolev spaces with respect to the probability measure μ, even if μ may be taken as a general measure. As expected, if μ is taken as the usual Lebesgue measure, we return to the standard Sobolev spaces.

So, for a probability measure μ, we denote by $W_\mu^{1,p}$ the space of those functions $\phi \in L^p(d\mu)$ for which there exist functions $\theta_i \in L^p(d\mu)$, $i = 1, \ldots, d$, such that $\int \phi \partial_i f d\mu = -\int \theta_i f d\mu$ for every test function $f \in C_c^\infty(\mathbb{R}^d)$. If F is a random variable of law μ, then the above duality relation reads $\mathbb{E}(\phi(F)\partial_i f(F)) = -\mathbb{E}(\theta_i(F)f(F))$ and these are the usual integration by parts formulas in the probabilistic approach; for example $-\theta_i(F)$ is connected to the weight produced by Malliavin calculus for a functional F on the Wiener space. But one may consider other frameworks, such as the Malliavin calculus on the Wiener Poisson space, for example. This formalism has already been used by Shigekawa in [45] and a slight variant appears in the book by Malliavin and Thalmaier [36] (the so-called divergence for covering vector fields).

In Section 1.1 we prove the following result: if $1 \in W_\mu^{1,p}$ (or equivalently $\mathrm{IBP}_i(F,1)$, $i = 1, \ldots, d$, holds) for some $p > d$, then

$$\sup_{x \in \mathbb{R}^d} \sum_{i=1}^d \int |\partial_i Q_d(y-x)|^{p/(p-1)} \mu(dy) \leq C_{d,p} \|1\|_{W_\mu^{1,p}}^{\ell_{d,p}},$$

where $\ell_{d,p} > 0$ denotes a suitable constant (see Theorem 1.4.1 for details). Moreover, $\mu(dx) = p_\mu(x)dx$, with p_μ Hölder continuous of order $1 - d/p$, and the following representation formula holds:

$$p_\mu(x) = \sum_{i=1}^d \int \partial_i Q_d(y-x) \partial_i^\mu 1(y) \mu(dy).$$

More generally, let $\mu_\phi(dx) := \phi(x)\mu(dx)$. If $\phi \in W_\mu^{1,p}$, then $\mu_\phi(dx) = p_{\mu_\phi}(x)dx$ and p_{μ_ϕ} is Hölder continuous. This last generalization is important from a probabilistic point of view because it produces a representation formula and regularity properties for the conditional expectation. We introduce in a straightforward way higher-order Sobolev spaces $W_\mu^{m,p}$, $m \in \mathbb{N}$, and we prove that if $1 \in W_\mu^{m,p}$, then p_μ is $m-1$ times differentiable and the derivatives of order $m-1$ are Hölder continuous. And the analogous result holds for $\phi \in W_\mu^{m,p}$. So if we are able to iterate m times the integration by parts formula we obtain a density in C^{m-1}. We also obtain some supplementary information about the queues of the density function and we develop more the applications allowing to handle the conditional expectations.

1.1 Sobolev spaces associated to probability measures

Let μ denote a probability measure on $(\mathbb{R}^d, \mathscr{B}(\mathbb{R}^d))$. We set

$$L_\mu^p = \left\{ \phi : \int |\phi(x)|^p \, \mu(dx) < \infty \right\}, \quad \text{with } \|\phi\|_{L_\mu^p} = \left(\int |\phi(x)|^p \, \mu(dx) \right)^{1/p}.$$

We also denote by $W_\mu^{1,p}$ the space of those functions $\phi \in L_\mu^p$ for which there exist functions $\theta_i \in L_\mu^p$, $i = 1, \ldots, d$, such that the following integration by parts formula holds:

$$\int \partial_i f(x)\phi(x)\mu(dx) = - \int f(x)\theta_i(x)\mu(dx) \quad \forall f \in C_c^\infty(\mathbb{R}^d).$$

We write $\partial_i^\mu \phi = \theta_i$ and we define the norm

$$\|\phi\|_{W_\mu^{1,p}} = \|\phi\|_{L_\mu^p} + \sum_{i=1}^d \|\partial_i^\mu \phi\|_{L_\mu^p}.$$

We similarly define higher-order Sobolev spaces. Let $k \in \mathbb{N}$ and $\alpha = (\alpha_1, \ldots, \alpha_k) \in \{1, \ldots, d\}^k$ be a multiindex. We denote by $|\alpha| = k$ the length of α and, for a function $f \in C^k(\mathbb{R}^d)$, we denote by $\partial_\alpha f = \partial_{\alpha_k} \cdots \partial_{\alpha_1} f$ the standard derivative corresponding to the multiindex α. Then we define $W_\mu^{m,p}$ to be the space of those functions $\phi \in L_\mu^p$ such that for every multiindex α with $|\alpha| = k \leq m$ there exist functions $\theta_\alpha \in L_\mu^p$ such that

$$\int \partial_\alpha f(x)\phi(x)\mu(dx) = (-1)^{|\alpha|} \int f(x)\theta_\alpha(x)\mu(dx) \quad \forall f \in C_c^\infty(\mathbb{R}^d).$$

We denote $\partial_\alpha^\mu \phi = \theta_\alpha$ and we define the norm

$$\|\phi\|_{W_\mu^{m,p}} = \sum_{0 \leq |\alpha| \leq m} \|\partial_\alpha^\mu \phi\|_{L_\mu^p},$$

with the understanding that $\partial_\alpha^\mu \phi = \phi$ when α is the empty multiindex (that is, $|\alpha| = 0$). It is easy to check that $(W_\mu^{m,p}, \|\cdot\|_{W_\mu^{m,p}})$ is a Banach space.

   It is worth remarking that in the above definitions it is not needed that μ be a probability measure: instead, a general measure μ can be used, not necessarily of mass 1 or even finite. Notice that one recovers the standard Sobolev spaces when μ is the Lebesgue measure. And in fact, we will use the notation $L^p, W^{m,p}$ for the spaces associated to the Lebesgue measure (instead of μ), which are the standard L^p and the standard Sobolev spaces used in the literature. If $D \subset \mathbb{R}^d$ is an open set, we denote by $W_\mu^{m,p}(D)$ the space of those functions ϕ which verify the integration by parts formula for test functions f with compact support included in D (so $W_\mu^{m,p} = W_\mu^{m,p}(\mathbb{R}^d)$). And similarly for $W^{m,p}(D), L^p(D), L_\mu^p(D)$.

   The "derivatives" ∂_α^μ defined on the space $W_\mu^{m,p}$ generalize the standard (weak) derivatives being, in fact, equal to them when μ is the Lebesgue measure. A nice difference consists in the fact that the function $\phi = 1$ does not necessarily belong to $W_\mu^{1,p}$. The following computational rules hold.

Proposition 1.1.1.   (i) *If $\phi \in W_\mu^{1,p}$ and $\psi \in C_b^1(\mathbb{R}^d)$, then $\psi\phi \in W_\mu^{1,p}$ and*

$$\partial_i^\mu(\psi\phi) = \psi\partial_i^\mu \phi + \partial_i\psi\phi; \tag{1.1}$$

in particular, if $1 \in W_\mu^{1,p}$, then for any $\psi \in C_b^1(\mathbb{R}^d)$ we have $\psi \in W_\mu^{1,p}$ and

$$\partial_i^\mu \psi = \psi\partial_i^\mu 1 + \partial_i\psi. \tag{1.2}$$

(ii) *Suppose that* $1 \in W_\mu^{1,p}$. *If* $\psi \in C_b^1(\mathbb{R}^m)$ *and if* $u = (u_1, \ldots, u_m) \colon \mathbb{R}^d \to \mathbb{R}^m$ *is such that* $u_j \in C_b^1(\mathbb{R}^d)$, $j = 1, \ldots, m$, *then* $\psi \circ u \in W_\mu^{1,p}$ *and*

$$\partial_i^\mu(\psi \circ u) = \sum_{j=1}^m (\partial_j \psi) \circ u \partial_i^\mu u_j + T_\psi \circ u \, \partial_i^\mu 1,$$

where

$$T_\psi(x) = \psi(x) - \sum_{j=1}^d \partial_j \psi(x) x_j.$$

Proof. (i) Since ψ and $\partial_i \psi$ are bounded, $\psi\phi, \psi\partial_i^\mu\phi, \partial_i\psi\phi \in L_\mu^p$. So we just have to check the integration by parts formula. We have

$$\int \partial_i f \psi\phi \, d\mu = \int \partial_i(f\psi)\phi \, d\mu - \int f\partial_i\psi\phi \, d\mu = -\int f\psi\partial_i^\mu \phi \, d\mu - \int f\partial_i\psi\phi \, d\mu$$

and the statement holds.

(ii) By using (1.2), we have

$$\partial_i^\mu(\psi \circ u) = \psi \circ u \, \partial_i^\mu 1 + \partial_i(\psi \circ u) = \psi \circ u \, \partial_i^\mu 1 + \sum_{j=1}^m (\partial_j\psi) \circ u \, \partial_i u_j.$$

The formula now follows by inserting $\partial_i u_j = \partial_i^\mu u_j - u_j \partial_i^\mu 1$, as given by (1.2). □

Let us recall that a similar formalism has been introduced by Shigekawa [45] and is close to the concept of "divergence for vector fields with respect to a probability measure" developed in Malliavin and Thalmaier [36, Sect. 4.2]. Moreover, also Bichteler, Gravereaux, and Jacod [16] have developed a variant of Sobolev's lemma, involving integration by parts type formulas in order to study the existence and regularity of the density (see Section 4 therein).

1.2 The Riesz transform

Our aim is to study the link between $W_\mu^{m,p}$ and $W^{m,p}$, our main tool being the Riesz transform, which is linked to the Poisson kernel Q_d, i.e., the solution to the equation $\Delta Q_d = \delta_0$ in $\mathbb{R}^d$ (δ_0 denoting the Dirac mass at 0). Here, Q_d has the following explicit form:

$$Q_1(x) = \max\{x, 0\}, \quad Q_2(x) = a_2^{-1} \ln|x| \quad \text{and} \quad Q_d(x) = -a_d^{-1}|x|^{2-d}, \ d > 2, \tag{1.3}$$

where a_d is the area of the unit sphere in $\mathbb{R}^d$. For $f \in C_c^1(\mathbb{R}^d)$ we have

$$f = (\nabla Q_d) * \nabla f, \tag{1.4}$$

the symbol $*$ denoting convolution. In Malliavin and Thalmaier [36, Theorem 4.22], the representation (1.4) for f is called the Riesz transform of f, and is employed in order to obtain representation formulas for the conditional expectation. Moreover, similar representation formulas for functions on the sphere and on the ball are used in Malliavin and Nualart [35] in order to obtain lower bounds for the density of a strongly non-degenerate random variable.

Let us discuss the integrability properties of the Poisson kernel. First, let $d \geq 2$. Setting $A_2 = 1$ and for $d > 2$, $A_d = d - 2$, we have

$$\partial_i Q_d(x) = a_d^{-1} A_d \frac{x_i}{|x|^d}. \tag{1.5}$$

By using polar coordinates, for $R > 0$ we obtain

$$\int_{|x| \leq R} \left| \partial_i Q_d(x) \right|^{1+\delta} dx \leq a_d^{-(1+\delta)} A_d^{1+\delta} a_d \int_0^R \left| \frac{r}{r^d} \right|^{1+\delta} r^{d-1} dr$$

$$= a_d^{-\delta} A_d^{1+\delta} \int_0^R \frac{1}{r^{\delta(d-1)}} dr, \tag{1.6}$$

which is finite for any $\delta < 1/(d-1)$. But $\left| \partial_i^2 Q_d(x) \right| \sim |x|^{-d}$ and so,

$$\int_{|x| \leq 1} \left| \partial_i^2 Q_d(x) \right| dx = \infty.$$

This is the reason why we have to integrate by parts once and to remove one derivative, but we may keep the other derivative.

On the other hand, again by using polar coordinates, we get

$$\int_{|x| > R} \left| \partial_i Q_d(x) \right|^{1+\delta} dx \leq a_d^{-\delta} A_d^{1+\delta} \int_R^{+\infty} \frac{1}{r^{\delta(d-1)}} dr, \tag{1.7}$$

and the above integral is finite when $\delta > 1/(d-1)$. So, the behaviors at 0 and ∞ are completely different: powers which are integrable around 0 may not be at ∞, and viceversa —one must take care of this fact.

In order to include the one-dimensional case, we notice that

$$\frac{dQ_1(x)}{dx} = 1_{(0,\infty)}(x),$$

so that powers of this function are always integrable around 0 and never at ∞.

Now, coming back to the representation (1.4), for every f whose support is included in $B_R(0)$ we deduce

$$\|f\|_\infty = \|\nabla Q_d * \nabla f\|_\infty \leq C_{R,p} \|\nabla f\|_p \tag{1.8}$$

for every $p > d$, so a Poincaré type inequality holds. In Theorem 1.4.1 below we will give an estimate of the Riesz transform providing a generalization of the above inequality to probability measures.

1.3 A first absolute continuity criterion: Malliavin–Thalmaier representation formula

For a function $\phi \in L^1_\mu$ we denote by μ_ϕ the signed finite measure defined by

$$\mu_\phi(dx) := \phi(x)\mu(dx).$$

We state now a first result on the existence and the regularity of the density which is the starting point for next results.

Theorem 1.3.1. (i) *Let* $\phi \in W^{1,1}_\mu$. *Then*

$$\int \left| \partial_i Q_d(y-x)\partial_i^\mu \phi(y)\right| \mu(dy) < \infty$$

for a.e. $x \in \mathbb{R}^d$, *and* $\mu_\phi(dx) = p_{\mu_\phi}(x)dx$ *with*

$$p_{\mu_\phi}(x) = -\sum_{i=1}^d \int \partial_i Q_d(y-x)\partial_i^\mu \phi(y)\mu(dy). \tag{1.9}$$

(ii) *If* $\phi \in W^{m,1}_\mu$ *for some* $m \geq 2$, *then* $p_{\mu_\phi} \in W^{m-1,1}$ *and, for each multiindex* α *of length less than or equal to* $m-1$, *the associated weak derivative can be represented as*

$$\partial_\alpha p_{\mu_\phi}(x) = -\sum_{i=1}^d \int \partial_i Q_d(y-x)\partial^\mu_{(\alpha,i)}\phi(y)\mu(dy). \tag{1.10}$$

If, in addition, $1 \in W^{1,1}_\mu$, *then the following alternative representation formula holds:*

$$\partial_\alpha p_{\mu_\phi}(x) = p_\mu(x)\partial^\mu_\alpha \phi(x); \tag{1.11}$$

in particular, taking $\phi = 1$ *and* $\alpha = \{i\}$ *we have*

$$\partial_i^\mu 1 = 1_{\{p_\mu > 0\}}\partial_i \ln p_\mu. \tag{1.12}$$

Remark 1.3.2. Representation (1.9) has been already proved in Malliavin and Thalmaier [36], though written in a slightly different form, using the "covering vector fields" approach. It is used in a probabilistic context in order to represent the conditional expectation. We will see this property in Section 1.8 below.

Remark 1.3.3. Formula (1.9) can be written as $p_{\mu_\phi} = -\nabla Q_d *_\mu \partial^\mu \phi$, where $*_\mu$ denotes the convolution with respect to the measure μ. So, there is an analogy with formula (1.4). Once we will be able to give an estimate for ∇Q_d in L^p_μ, we will state a Poincaré type estimate similar to (1.8).

Proof of Theorem 1.3.1. (i) We take $f \in C_c^1(\mathbb{R}^d)$ and we write $f = \Delta(Q_d * f) = \sum_{i=1}^d (\partial_i Q_d) * (\partial_i f)$. Then

$$
\int f d\mu_\phi = \int f\phi d\mu = \sum_{i=1}^d \int \mu(dx)\phi(x) \int \partial_i Q_d(z)\partial_i f(x-z)dz
$$

$$
= \sum_{i=1}^d \int \partial_i Q_d(z) \int \mu(dx)\phi(x)\partial_i f(x-z))dz
$$

$$
= -\sum_{i=1}^d \int \partial_i Q_d(z) \int \mu(dx)f(x-z)\partial_i^\mu \phi(x)dz
$$

$$
= -\sum_{i=1}^d \int \mu(dx)\partial_i^\mu \phi(x) \int \partial_i Q_d(z)f(x-z)dz
$$

$$
= -\sum_{i=1}^d \int \mu(dx)\partial_i^\mu \phi(x) \int \partial_i Q_d(x-y)f(y)dy
$$

$$
= \int f(y)\left(-\sum_{i=1}^d \int \partial_i Q_d(x-y)\partial_i^\mu \phi(x)\mu(dx) \right)dy,
$$

which proves the representation formula (1.9).

   In the previous computations we have used several times Fubini's theorem, so we need to prove that some integrability properties hold. Let us suppose that the support of f is included in $B_R(0)$ for some $R > 1$. We denote $C_R(x) = \{y : |x| - R \le |y| \le |x| + R\}$, and then $B_R(x) \subset C_R(x)$. First of all

$$
|\phi(x)\partial_i Q_d(z)\partial_i f(x-z)| \le \|\partial_i f\|_\infty |\phi(x)| |\partial_i Q_d(z)1_{C_R(x)}(z)|
$$

and

$$
\int_{C_R(x)} |\partial_i Q_d(z)|dz \le A_d \int_{|x|-R}^{|x|+R} \frac{r}{r^d} \times r^{d-1}dr = 2RA_d.
$$

Therefore,

$$
\int\int |\phi(x)\partial_i Q_d(z)\partial_i f(x-z)| \, dz\mu(dx) \le 2RA_d\|\partial_i f\|_\infty \int |\phi(x)| \, \mu(dx) < \infty.
$$

Similarly,

$$
\int\int |\partial_i^\mu \phi(x)\partial_i Q_d(z)f(x-z)|dz\mu(dx) = \int\int |\partial_i^\mu \phi(x)\partial_i Q_d(x-y)f(y)|dy\mu(dx)
$$

$$
\le 2RA_d\|f\|_\infty \int |\partial_i^\mu \phi(x)| \, \mu(dx) < \infty
$$

and so, all the needed integrability properties hold and our computations are correct. In particular, we have checked that $\int dy f(y) \int |\partial_i^\mu \phi(x)\partial_i Q_d(x-y)| \, \mu(dx) <$

∞ for every $f \in C_c^1(\mathbb{R}^d)$, and so $\int |\partial_i^\mu \phi(x) \partial_i Q_d(x-y)| \, \mu(dx)$ is finite dy almost surely.

(ii) In order to prove (1.10), we write $\partial_\alpha f = \sum_{i=1}^d \partial_i Q_d * \partial_i \partial_\alpha f$. Now, we use the same chain of equalities as above and we obtain

$$\int \partial_\alpha f(x) p_{\mu_\phi}(x) dx = \int \partial_\alpha f d\mu_\phi$$

$$= (-1)^{|\alpha|} \int f(y) \left(-\sum_{i=1}^d \int \partial_i Q_d(x-y) \partial_{(\alpha,i)}^\mu \phi(x) \mu(dx) \right) dy,$$

so that $\partial_\alpha p_{\mu_\phi}(y) = -\sum_{i=1}^d \int \partial_i Q_d(x-y) \partial_{(\alpha,i)}^\mu \phi(x) \mu(dx)$. As for (1.11), we have

$$\int \partial_\alpha f(x) p_{\mu_\phi}(x) dx = \int \partial_\alpha f(x) \mu_\phi(dx) = \int \partial_\alpha f(x) \phi(x) \mu(dx)$$

$$= (-1)^{|\alpha|} \int f(x) \partial_\alpha^\mu \phi(x) \mu(dx)$$

$$= (-1)^{|\alpha|} \int f(x) \partial_\alpha^\mu \phi(x) p_\mu(x) dx. \qquad \square$$

Remark 1.3.4. Notice that if $1 \in W_\mu^{1,1}$, then for any $f \in C_c^\infty$ we have

$$\left| \int \partial_i f d\mu \right| \le c_i \|f\|_\infty, \quad \text{with } c_i = \|\partial_i^\mu 1\|_{L_\mu^1}, \quad i = 1, \dots, d.$$

Now, it is known that the above condition implies the existence of the density, as proved by Malliavin in [33] (see also D. Nualart [41, Lemma 2.1.1]), and Theorem 1.3.1 gives a new proof including the representation formula in terms of the Riesz transform.

1.4 Estimate of the Riesz transform

As we will see later, it is important to be able to control ∇Q_d under the probability measure μ. More precisely, we consider the quantity $\Theta_p(\mu)$ defined by

$$\Theta_p(\mu) = \sup_{a \in \mathbb{R}^d} \sum_{i=1}^d \left(\int_{\mathbb{R}^d} \left| \partial_i Q_d(x-a) \right|^{\frac{p}{p-1}} \mu(dx) \right)^{\frac{p-1}{p}}, \qquad (1.13)$$

that is, $\Theta_p(\mu) = \sup_{a \in \mathbb{R}^d} \sum_{i=1}^d \|\partial_i Q_d(\cdot - a)\|_{L_\mu^q}$, where q is the conjugate of p. The main result of this section is the following theorem.

Theorem 1.4.1. *Let $p > d$ and let μ be a probability measure on $\mathbb{R}^d$ such that $1 \in W_\mu^{1,p}$. Then we have*

$$\Theta_p(\mu) \le 2dK_{d,p} \|1\|_{W_\mu^{1,p}}^{k_{d,p}}, \qquad (1.14)$$

where

$$k_{d,p} = \frac{(d-1)p}{p-d} \quad and \quad K_{d,p} = 1 + 2(a_d^{-1}A_d)^{\frac{p}{p-1}}\left(\frac{p-1}{p-d}\cdot 2d\,(a_d^{-1}A_d)^{\frac{p}{p-1}}a_d\right)^{k_{d,p}}.$$

Remark 1.4.2. The inequality (1.14) gives an estimate of the kernels $\partial_i Q_d$, $i = 1,\ldots,d$, which appear in the Riesz transform; this is the crucial point in our approach. In Malliavin and Nualart [35], the authors use the Riesz transform on the sphere and they give estimates of the L^p norms of the corresponding kernels (which are of course different).

Remark 1.4.3. Under the hypotheses of Theorem 1.4.1, by applying Theorem 1.3.1 we get $\mu(dx) = p_\mu(x)dx$ with p_μ given by

$$p_\mu(x) = -\sum_{i=1}^{d}\int \partial_i Q_d(y-x)\partial_i^\mu 1(y)\mu(dy). \tag{1.15}$$

Moreover, in the proof of Theorem 1.4.1 we prove that p_μ is bounded and

$$\|p_\mu\|_\infty \le 2dK_{d,p}\|1\|_{W_\mu^{1,p}}^{k_{d,p}+1}$$

(for details, see (1.17) below).

The proof of Theorem 1.4.1 needs two preparatory lemmas.

For a probability measure μ and a probability density ψ (a nonnegative measurable function ψ with $\int_{\mathbb{R}^d}\psi(x)dx = 1$) we define the probability measure $\psi * \mu$ by

$$\int f(x)(\psi * \mu)(dx) := \int \psi(x)\int f(x+y)\mu(dy)dx.$$

Lemma 1.4.4. *Let $p \ge 1$. If $1 \in W_\mu^{1,p}$, then $1 \in W_{\psi*\mu}^{1,p}$ and $\|1\|_{W_{\psi*\mu}^{1,p}} \le \|1\|_{W_\mu^{1,p}}$, for every probability density ψ.*

Proof. On a probability space $(\Omega, \mathcal{F}, P)$ we consider two independent random variables F and Δ such that $F \sim \mu(dx)$ and $\Delta \sim \psi(x)dx$. Then $F + \Delta \sim (\psi * \mu)(dx)$. We define $\theta_i(x) = \mathbb{E}(\partial_i^\mu 1(F) \mid F + \Delta = x)$ and we claim that $\partial_i^{\psi*\mu}1 = \theta_i$. In fact, for $f \in C_c^1(\mathbb{R}^d)$ we have

$$-\int \partial_i f(x)(\psi * \mu)(dx) = -\int dx\psi(x)\int \partial_i f(x+y)\mu(dy)$$

$$= -\int dx\psi(x)\int f(x+y)\partial_i^\mu 1(y)\mu(dy)$$

$$= \mathbb{E}\big(f(F+\Delta)\partial_i^\mu 1(F)\big)$$

$$= \mathbb{E}\big(f(F+\Delta)\mathbb{E}\big(\partial_i^\mu 1(F) \mid F+\Delta\big)\big)$$

$$= \mathbb{E}\big(f(F+\Delta)\theta_i(F+\Delta)\big)$$

$$= \int f(x) \times \theta_i(x)(\psi * \mu)(dx),$$

so $\partial_i^{\psi * \mu} 1 = \theta_i$. Moreover,

$$\int |\theta_i(x)|^p \, (\psi * \mu)(dx) = \mathbb{E}\big(|\theta_i(F + \Delta)|^p\big) = \mathbb{E}\big(|\mathbb{E}(\partial_i^\mu 1(F) \mid F + \Delta)|^p\big)$$

$$\leq \mathbb{E}\big(|\partial_i^\mu 1(F)|^p\big) = \int |\partial_i^\mu 1(x)|^p \, \mu(dx),$$

so $1 \in W_{\psi * \mu}^{1,p}$ and $\|1\|_{W_{\psi * \mu}^{1,p}} \leq \|1\|_{W_\mu^{1,p}}$. $\qquad\square$

Lemma 1.4.5. *Let p_n, $n \in \mathbb{N}$, be a sequence of probability densities such that*

$$\sup_n \|p_n\|_\infty = C_\infty < \infty.$$

Suppose also that the sequence of probability measures $\mu_n(dx) = p_n(x)dx$, $n \in \mathbb{N}$, converges weakly to a probability measure μ. Then, $\mu(dx) = p(x)dx$ and $\|p\|_\infty \leq C_\infty$.

Proof. Since $\int p_n^2(x)dx \leq C_\infty$, the sequence p_n is bounded in $L^2(\mathbb{R}^d)$ and so it is weakly relatively compact. Passing to a subsequence (which we still denote by p_n) we can find $p \in L^2(\mathbb{R}^d)$ such that $\int p_n(x)f(x)dx \to \int p(x)f(x)dx$ for every $f \in L^2(\mathbb{R}^d)$. But, if $f \in C_c(\mathbb{R}^d) \subset L^2(\mathbb{R}^d)$, then $\int p_n(x)f(x)dx \to \int f(x)\mu(dx)$, so that $\mu(dx) = p(x)dx$.

Let us now check that p is bounded. Using Mazur's theorem, we can construct a convex combination $q_n = \sum_{i=1}^{m_n} \lambda_i^n p_{n+i}$, with $\lambda_i^n \geq 0$, $\sum_{i=1}^{m_n} \lambda_i^n = 1$, such that $q_n \to p$ strongly in $L^2(\mathbb{R}^d)$. Then, passing to a subsequence, we may assume that $q_n \to p$ almost everywhere. It follows that $p(x) \leq \sup_n q_n(x) \leq C_\infty$ almost everywhere. And we can change p on a set of null measure. $\qquad\square$

We are now able to prove Theorem 1.4.1.

Proof of Theorem 1.4.1. We recall Remark 1.4.3: we know that the probability density p_μ exists and representation (1.15) holds. So, we first prove the theorem under the supplementary assumption:

(H) $\qquad\qquad\qquad\qquad p_\mu$ is bounded.

We take $\rho > 0$ and note that if $|x - a| > \rho$, then $|\partial_i Q_d(x - a)| \leq a_d^{-1} A_d \rho^{-(d-1)}$. Since $p > d$, for any $a \in \mathbb{R}^d$ we have

$$\int |\partial_i Q_d(x - a)|^{\frac{p}{p-1}} \mu(dx) \leq B_d^{\frac{p}{p-1}} \rho^{-(d-1)\frac{p}{p-1}} + \int_{|x-a| \leq \rho} |\partial_i Q_d(x - a)|^{\frac{p}{p-1}} p_\mu(x) \, dx$$

$$\leq B_d^{\frac{p}{p-1}} \left[\rho^{-(d-1)\frac{p}{p-1}} + \|p_\mu\|_\infty a_d \int_0^\rho \frac{dr}{r^{\frac{d-1}{p-1}}} \right],$$

in which we have set $B_d = a_d^{-1} A_d$. This gives

$$\int |\partial_i Q_d(x-a)|^{\frac{p}{p-1}} \mu(dx) \le B_d^{\frac{p}{p-1}} \left[\rho^{-(d-1)\frac{p}{p-1}} + a_d \|p_\mu\|_\infty \frac{p-1}{p-d} \rho^{\frac{p-d}{p-1}} \right] < \infty. \tag{1.16}$$

Using the representation formula (1.15) and Hölder's inequality, we get

$$p_\mu(x) = - \sum_{i=1}^d \int \partial_i Q_d(y-x) \partial_i^\mu 1(y) \mu(dy)$$

$$\le \|1\|_{W_\mu^{1,p}} \sum_{i=1}^d \left(\int |\partial_i Q_d(y-x)|^{\frac{p}{p-1}} \mu(dy) \right)^{\frac{p-1}{p}}$$

$$\le \|1\|_{W_\mu^{1,p}} \left(d + \sum_{i=1}^d \int |\partial_i Q_d(y-x)|^{\frac{p}{p-1}} \mu(dy) \right). \tag{1.17}$$

By using (1.16), we obtain

$$\|p_\mu\|_\infty \le d \, \|1\|_{W_\mu^{1,p}} \left[B_d^{\frac{p}{p-1}} \left(\rho^{-(d-1)\frac{p}{p-1}} + a_d \|p_\mu\|_\infty \frac{p-1}{p-d} \rho^{\frac{p-d}{p-1}} \right) + 1 \right].$$

Choose now $\rho = \rho_*$, with ρ_* such that

$$d \, B_d^{\frac{p}{p-1}} a_d \, \|1\|_{W_\mu^{1,p}} \frac{p-1}{p-d} \rho_*^{\frac{p-d}{p-1}} = \frac{1}{2}$$

that is,

$$\rho_* = \left(\frac{p-1}{p-d} \cdot 2d \, B_d^{\frac{p}{p-1}} a_d \, \|1\|_{W_\mu^{1,p}} \right)^{-\frac{p-1}{p-d}}.$$

Then,

$$\|p_\mu\|_\infty \le 2d \, \|1\|_{W_\mu^{1,p}} \left(B_d^{\frac{p}{p-1}} \rho_*^{-(d-1)\frac{p}{p-1}} + 1 \right).$$

Since $\frac{p-1}{p-d} \rho_*^{\frac{p-d}{p-1}} = (2d B_d^{p/(p-1)} a_d \, \|1\|_{W_\mu^{1,p}})^{-1}$, by using (1.16) we obtain

$$\int |\partial_i Q_d(x-a)|^{\frac{p}{p-1}} \mu(dx) \le 1 + 2 B_d^{\frac{p}{p-1}} \rho_*^{-(d-1)\frac{p}{p-1}}$$

$$= 1 + 2 B_d^{\frac{p}{p-1}} \left(\frac{p-1}{p-d} \cdot 2d \, B_d^{\frac{p}{p-1}} a_d \right)^{\frac{(d-1)p}{p-d}} \cdot \|1\|_{W_\mu^{1,p}}^{\frac{(d-1)p}{p-d}}$$

$$\le \left[1 + 2 B_d^{\frac{p}{p-1}} \left(\frac{p-1}{p-d} \cdot 2d \, B_d^{\frac{p}{p-1}} a_d \right)^{\frac{(d-1)p}{p-d}} \right] \|1\|_{W_\mu^{1,p}}^{k_{d,p}}$$

$$= K_{d,p} \cdot \|1\|_{W_\mu^{1,p}}^{k_{d,p}}.$$

Finally,

$$\sum_{i=1}^{d} \int \left| \partial_i Q_d(y-x) \right|^{\frac{p}{p-1}} \mu(dy) \leq d\left(1+K_{d,p}\right) \|1\|_{W_\mu^{1,p}}^{k_{d,p}} \leq 2dK_{d,p} \|1\|_{W_\mu^{1,p}}^{k_{d,p}}.$$

Notice that, by using (1.17), we get

$$\|p_\mu\|_\infty \leq 2dK_{d,p} \|1\|_{W_\mu^{1,p}}^{k_{d,p}+1},$$

as already mentioned in Remark 1.4.3.

So the theorem is proved under the supplementary assumption (H). To remove this assumption, consider a nonnegative and continuous function ψ such that $\int \psi = 1$ and $\psi(x) = 0$ for $|x| \geq 1$. Then we define $\psi_n(x) = n^d \psi(nx)$ and $\mu_n = \psi_n * \mu$. We have $\mu_n(dx) = p_n(x)dx$ with $p_n(x) = \int \psi_n(x-y)\mu(dy)$. Using Lemma 1.4.4 we have $1 \in W_{\mu_n}^{1,p}$ and $\|1\|_{W_{\mu_n}^{1,p}} \leq \|1\|_{W_\mu^{1,p}} < \infty$. Since p_n is bounded, μ_n verifies assumption (H) and so, using the first part of the proof, we obtain

$$\|p_n\|_\infty \leq 2dK_{d,p} \|1\|_{W_{\mu_n}^{1,p}}^{k_{d,p}+1} \leq 2dK_{d,p} \|1\|_{W_\mu^{1,p}}^{k_{d,p}+1}.$$

Clearly $\mu_n \to \mu$ weakly so, using Lemma 1.4.5, we can find p such that $\mu(dx) = p(x)dx$ and p is bounded. So μ itself satisfies (H) and the proof is completed. $\square$

1.5 Regularity of the density

Theorem 1.3.1 says that μ_ϕ has a density as soon as $\phi \in W_\mu^{1,1}$ —and this does not depend on the dimension d of the space. But if we want to obtain a continuous or a differentiable density, we need more regularity for ϕ. The main tool for obtaining such properties is the classical theorem of Morrey, which we recall now (see Brezis [18, Corollary IX.13]).

Theorem 1.5.1. *Let $u \in W^{1,p}(\mathbb{R}^d)$. If $1 - d/p > 0$, then u is Hölder continuous of exponent $q = 1 - d/p$. Furthermore, suppose that $u \in W^{m,p}(\mathbb{R}^d)$ and $m - d/p > 0$. Let $k = [m - d/p]$ be the integer part of $m - d/p$, and $q = \{m - d/p\}$ its fractional part. If $k = 0$, then u is Hölder continuous of exponent q, and if $k \geq 1$, then $u \in C^k$ and, for any multiindex α with $|\alpha| \leq k$, the derivative $\partial_\alpha u$ is Hölder continuous of exponent q, i.e., for any $x, y \in \mathbb{R}^d$,*

$$\left| \partial_\alpha u(x) - \partial_\alpha u(y) \right| \leq C_{d,p} \|u\|_{W^{m,p}(\mathbb{R}^d)} |x-y|^q,$$

where $C_{d,p}$ depends only on d and p.

It is clear from Theorem 1.5.1 that there are two ways to improve the regularity of u: we have to increase m or/and p. If $\phi \in W_\mu^{m,p}$ for a sufficiently large m, then Theorem 1.3.1 already gives us a differentiable density p_{μ_ϕ}. But if we want to keep m low we have to increase p. And, in order to be able to do it, the key point is the estimate for $\Theta_p(\mu)$ given in Theorem 1.4.1.

Theorem 1.5.2. *Let $p > d$ and $1 \in W_\mu^{1,p}$. For $m \geq 1$, let $\phi \in W_\mu^{m,p}$, so that $\mu_\phi(dx) = p_{\mu_\phi}(x)dx$. Then, $p_{\mu_\phi} \in W^{m,p}$. As a consequence, $p_{\mu_\phi} \in C^{m-1}$ and*

$$\left\|p_{\mu_\phi}\right\|_{W^{m,p}} \leq (2dK_{d,p})^{1-1/p} \|1\|_{W_\mu^{1,p}}^{k_{d,p}(1-1/p)} \|\phi\|_{W_\mu^{m,p}}. \tag{1.18}$$

Moreover, for any multiindex α such that $0 \leq |\alpha| = \ell \leq m - 1$,

$$\left\|\partial_\alpha p_{\mu_\phi}\right\|_\infty \leq dK_{d,p} \|1\|_{W_\mu^{1,p}}^{k_{d,p}} \|\phi\|_{W_\mu^{\ell+1,p}}. \tag{1.19}$$

Remark 1.5.3. For $\alpha = \emptyset$, (1.19) reads

$$\left\|p_{\mu_\phi}\right\|_\infty \leq dK_{d,p} \|1\|_{W_\mu^{1,p}}^{k_{d,p}} \|\phi\|_{W_\mu^{1,p}}.$$

Actually, a closer look at the proof will give

$$\left\|p_{\mu_\phi}\right\|_\infty \leq dK_{d,p} \|1\|_{W_\mu^{1,p}}^{k_{d,p}} \|\partial^\mu \phi\|_{L_\mu^p}$$

so that, in the notations used in Remark 1.3.3,

$$\left\|p_{\mu_\phi}\right\|_\infty = \|\nabla Q_d *_\mu \partial^\mu \phi\|_\infty \leq C_{p,d} \|\partial^\mu \phi\|_{L_\mu^p}$$

for all $p > d$. This can be considered as the analog of the Poincaré inequality (1.8) for probability measures.

Proof of Theorem 1.5.2. We use (1.11) (with the notation $\partial_\alpha^\mu \phi := \phi$ if $\alpha = \emptyset$) and obtain

$$\int \left|\partial_\alpha p_{\mu_\phi}(x)\right|^p dx = \int |\partial_\alpha^\mu \phi(x)|^p |p_\mu(x)|^p dx \leq \|p_\mu\|_\infty^{p-1} \int |\partial_\alpha^\mu \phi(x)|^p p_\mu(x) dx.$$

So, $\left\|\partial_\alpha p_{\mu_\phi}\right\|_{L^p} \leq \|p_\mu\|_\infty^{1-1/p} \|\partial_\alpha^\mu \phi\|_{L_\mu^p}$ and, by using (1.14), we obtain (1.18). Now, using the representation formula (1.10), Hölder's inequality, and Theorem 1.4.1 we get

$$\left|\partial_\alpha p_{\mu_\phi}(x)\right| \leq \Theta_p(\mu) \sum_{i=1}^d \|\partial_{(\alpha,i)}^\mu \phi\|_{L_\mu^p} \leq dK_{d,p} \|1\|_{W_\mu^{1,p}}^{k_{d,p}} \|\phi\|_{W_\mu^{\ell+1,p}},$$

and (1.19) is proved. The fact that $p_{\mu_\phi} \in C^{m-1}$ follows from Theorem 1.5.1. □

Remark 1.5.4. Under the hypotheses of Theorem 1.5.2, one has the following consequences:

(i) For any multiindex β such that $0 \leq |\beta| = k \leq m - 2$, $\partial_\beta p_{\mu_\phi}$ is Lipschitz continuous, i.e., for any $x, y \in \mathbb{R}^d$,

$$\left|\partial_\beta p_{\mu_\phi}(x) - \partial_\beta p_{\mu_\phi}(y)\right| \leq d^2 K_{d,p} \|1\|_{W_\mu^{1,p}}^{k_{d,p}} \|\phi\|_{W_\mu^{k+2,p}} |x - y|.$$

In fact, from (1.19) and the fact that if $f \in C^1$ with $\|\nabla f\|_\infty < \infty$, then $|f(x) - f(y)| \leq \sum_{i=1}^d \|\partial_i f\|_\infty |x - y|$, the statement follows immediately.

(ii) For any multiindex β such that $|\beta| = m - 1$, $\partial_\beta p_{\mu_\phi}$ is Hölder continuous of exponent $1 - d/p$, i.e., for any $x, y \in \mathbb{R}^d$,

$$\left| \partial_\beta p_{\mu_\phi}(x) - \partial_\beta p_{\mu_\phi}(y) \right| \leq C_{d,p} \|p_{\mu_\phi}\|_{W^{m,p}} |x - y|^{1-d/p},$$

where $C_{d,p}$ depends only on d and p. (This is a standard consequence of $p_{\mu_\phi} \in W^{m,p}$, as stated in Theorem 1.5.1.)

(iii) One has $W^{m,p}_\mu \subset \bigcap_{\delta > 0} W^{m,p}(\{p_\mu > \delta\})$. In fact, $p_{\mu_\phi}(x)dx = \phi(x)\mu(dx) = \phi(x)p_\mu(x)dx$, so $\phi(x) = p_{\mu_\phi}(x)/p_\mu(x)$ if $p_\mu(x) > 0$. And since $p_{\mu_\phi}, p_\mu \in W^{m,p}(\mathbb{R}^d)$, we obtain $\phi \in W^{m,p}(\{p_\mu > \delta\})$. $\qquad\square$

1.6 Estimate of the tails of the density

We give now a result which allows to estimate the tails of p_μ.

Proposition 1.6.1. *Let $\phi \in C^1_b(\mathbb{R}^d)$ be a function such that $1_{B_1(0)} \leq \phi \leq 1_{B_2(0)}$. Set $\phi_x(y) = \phi(x - y)$ and assume that $1 \in W^{1,p}_\mu$ with $p > d$. Then, $\phi_x \in W^{1,p}_\mu$ and the following representation holds:*

$$p_\mu(x) = \sum_{i=1}^d \int \partial_i Q_d(y - x) \partial_i^\mu \phi_x(y) 1_{\{|y-x|<2\}} \mu(dy).$$

As a consequence, for any positive $a < 1/d - 1/p$,

$$p_\mu(x) \leq \Theta_{\overline{p}}(\mu)\left(d + \|1\|_{W^{1,p}_\mu}\right) \mu\left(B_2(x)\right)^a, \tag{1.20}$$

where $\overline{p} = 1/(a + 1/p)$. In particular,

$$\lim_{|x| \to \infty} p_\mu(x) = 0. \tag{1.21}$$

Proof. By Proposition 1.1.1, $\phi_x \in W^{1,p}_\mu$ and $\partial_i^\mu \phi_x = \phi_x \partial_i^\mu 1 + \partial_i \phi_x$, so that $\partial_i^\mu \phi_x = \partial_i^\mu \phi_x 1_{B_2(x)}$. Now, for $f \in C^1_c(B_1(x))$, we have

$$\int f(y)\mu(dy) = \int f(y)\phi_x(y)\mu(dy) = \int f(y)p_{\mu_{\phi_x}}(y)dy$$

$$= -\int f(y) \sum_{i=1}^d \int \partial_i Q_d(z - y) \partial_i^\mu \phi_x(z) \mu(dz)dy$$

$$= -\int f(y) \sum_{i=1}^d \int \partial_i Q_d(z - y) \partial_i^\mu \phi_x(z) 1_{B_2(x)}(z) \mu(dz)dy.$$

It follows that for $y \in B_1(x)$ we have

$$p_\mu(y) = -\sum_{i=1}^d \int \partial_i Q_d(z - y) \partial_i^\mu \phi_x(z) 1_{B_2(x)}(z) \mu(dz).$$

Let now $y = x$ and take $a \in (0, 1/d - 1/p)$. Using Hölder's inequality, we obtain

$$p_\mu(x) \le \mu(B_2(x))^a \sum_{i=1}^d I_i \quad \text{with } I_i = \left(\int |\partial_i Q_d(z - x) \partial_i^\mu \phi_x(z)|^{\frac{1}{1-a}} \mu(dz) \right)^{1-a}.$$

Notice that $1 < d(1-a)/(1-da) < p(1-a)$. We take β such that $d(1-a)/(1-da) < \beta < p(1-a)$ and we denote by α the conjugate of β. Using again Hölder's inequality, we obtain

$$I_i \le \left(\int |\partial_i Q_d(z - x)|^{\frac{\alpha}{1-a}} \mu(dz) \right)^{(1-a)/\alpha} \left(\int |\partial_i^\mu \phi_x(z)|^{\frac{\beta}{1-a}} \mu(dz) \right)^{(1-a)/\beta}$$

$$\le \left(\int |\partial_i Q_d(z - x)|^{\frac{\alpha}{1-a}} \mu(dz) \right)^{(1-a)/\alpha} \left\| \partial_i^\mu \phi_x \right\|_{L_\mu^p}.$$

We let $\beta \uparrow p(1 - a)$ so that

$$\frac{\alpha}{1 - a} = \frac{\beta}{(\beta - 1)(1 - a)} \longrightarrow \frac{p}{p(1 - a) - 1} = \frac{\overline{p}}{\overline{p} - 1}.$$

Then we get

$$I_i \le \Theta_{\overline{p}}(\mu) \left\| \partial_i^\mu \phi_x \right\|_{L_\mu^p},$$

and so

$$p_\mu(x) \le \Theta_{\overline{p}}(\mu) \left\| \phi_x \right\|_{W_\mu^{1,p}} \mu(B_2(x))^a.$$

Now, since $\partial_i^\mu \phi_x = \phi_x \partial_i^\mu 1 + \partial_i \phi_x$, we have $\left\| \phi_x \right\|_{W_\mu^{1,p}} \le c_\phi d + \left\| 1 \right\|_{W_\mu^{1,p}}$, where $c_\phi > 1$ is such that $\left\| \nabla \phi \right\|_\infty \le c_\phi$, so that

$$p_\mu(x) \le \Theta_{\overline{p}}(\mu) \left(c_\phi d + \left\| 1 \right\|_{W_\mu^{1,p}} \right) \mu(B_2(x))^a.$$

Inequality (1.20) now follows by taking a sequence $\{\phi_n\}_n \subset C_b^1$ such that $1_{B_1(0)} \le \phi_n \le 1_{B_2(0)}$ with $\left\| \nabla \phi_n \right\|_\infty \le 1 + 1/n$, and by letting $n \to \infty$.

   Finally, $\mathbf{1}_{B_2(x)} \to 0$ a.s. when $|x| \to \infty$, and now the Lebesgue dominated convergence theorem shows that $\mu(B_2(x)) = \int \mathbf{1}_{B_2(x)}(y)\mu(dy) \to 0$. Applying (1.20), we obtain (1.21). $\qquad\square$

Remark 1.6.2. One can get a representation of the derivatives of the density function in a "localized" version as in Proposition 1.6.1, giving rise to estimates for the tails of the derivatives. In fact, take $\phi \in C_b^m(\mathbb{R}^d)$ to be a function such that $1_{B_1(0)} \le \phi \le 1_{B_2(0)}$, and set $\phi_x(y) = \phi(x - y)$. If $1 \in W_\mu^{m,p}$ with $p > d$, then it is clear that $\phi_x \in W_\mu^{m,p}$. And following the same arguments as in the previous proof, for a multiindex α with $|\alpha| \le m - 1$ we have

$$\partial_\alpha p_\mu(x) = (-1)^{|\alpha|} \sum_{i=1}^d \int \partial_i Q_d(y - x) \partial_{(i,\alpha)}^\mu \phi_x(y) 1_{\{|y-x|<2\}} \mu(dy).$$

1.7 Local integration by parts formulas and local densities

The global assumptions that have been made up to now may fail in concrete and interesting cases —for example, for diffusion processes living in a region of the space or, as a more elementary example, for the exponential distribution. So in this section we give a hint about the localized version of the results presented above.

Let D denote an open domain in $\mathbb{R}^d$. We recall that

$$L^p_\mu(D) = \left\{ f : \int_D |f(x)|^p d\mu(x) < \infty \right\}$$

and $W^{1,p}_\mu(D)$ is the space of those functions $\phi \in L^p_\mu(D)$ verifying the integration by parts formula $\int \phi \partial_i f d\mu = - \int \theta_i f d\mu$ for test functions f which have a compact support included in D, and $\partial_i^{\mu,D} \phi := \theta_i \in L^p_\mu(D)$. The space $W^{m,p}_\mu(D)$ is defined similarly. Our aim is to give sufficient conditions for μ to have a smooth density on D; this means that we look for a smooth function p_μ such that $\int_D f(x) d\mu(x) = \int_D f(x) p_\mu(x) dx$. And we want to give estimates for p_μ and its derivatives in terms of the Sobolev norms of $W^{m,p}_\mu(D)$.

The main step in our approach is the following truncation argument. Given $-\infty \le a \le b \le \infty$ and $\varepsilon > 0$, we define $\psi_{\varepsilon,a,b} : \mathbb{R} \to \mathbb{R}_+$ by

$$\psi_{\varepsilon,a,b}(x) = 1_{(a-\varepsilon,a]}(x) \exp\left(1 - \frac{\varepsilon^2}{\varepsilon^2 - (x-a)^2}\right) + 1_{(a,b)}(x)$$

$$+ 1_{[b,b+\varepsilon)}(x) \exp\left(1 - \frac{\varepsilon^2}{\varepsilon^2 - (x-b)^2}\right),$$

with the convention $1_{(a-\varepsilon,a)} = 0$ if $a = -\infty$ and $1_{(b,b+\varepsilon)} = 0$ if $b = \infty$. Notice that $\psi_{\varepsilon,a,b} \in C^\infty_b(\mathbb{R})$ and

$$\sup_{x \in (a-\varepsilon,b+\varepsilon)} \left|\partial_x \ln \psi_{\varepsilon,a,b}(x)\right|^p \psi_{\varepsilon,a,b}(x) \le 2^{p+1} \varepsilon^{-p} \sup_{y>0} y^{2p} e^{1-y}.$$

For $x = (x_1, \ldots, x_d)$ and $i \in \{1, \ldots, d\}$ we denote $\hat{x}^{(-i)} = (x_1, \ldots, x_{i-1}, x_{i+1}, \ldots, x_d)$ and, for $y \in \mathbb{R}$, we put $(\hat{x}^{(-i)}, y) = (x_1, \ldots, x_{i-1}, y, x_{i+1}, \ldots, x_d)$. Then, we define

$$a(\hat{x}^{(-i)}) = \inf_{y \in \mathbb{R}} \left\{ y : d\big((\hat{x}^{(-i)}, y), D^c\big) > 2\varepsilon \right\},$$

$$b(\hat{x}^{(-i)}) = \sup_{y \in \mathbb{R}} \left\{ y : d\big((\hat{x}^{(-i)}, y), D^c\big) > 2\varepsilon \right\},$$

with the convention $a(\hat{x}^{(-i)}) = b(\hat{x}^{(-i)}) = 0$ if $\{y : d((\hat{x}^{(-i)}, y), D^c) > 2\varepsilon\} = \emptyset$. Finally, we define $\Psi_{D,\varepsilon}(x) = \prod_{i=1}^d \psi_{\varepsilon,a(\hat{x}^{(-i)}),b(\hat{x}^{(-i)})}(x_i)$. We denote $D_\varepsilon = \{x : d(x, D^c) \ge \varepsilon\}$, so that $1_{D_{2\varepsilon}} \le \Psi_{D,\varepsilon} \le 1_{D_\varepsilon}$. And we also have

$$\sup_{x \in D_\varepsilon} \left|\partial_x \ln \Psi_{D,\varepsilon}(x)\right|^p \Psi_{D,\varepsilon}(x) \le d \, 2^{p+1} \varepsilon^{-p} \sup_{y>0} y^{2p} e^{1-y}. \tag{1.22}$$

We can now state the following preparatory lemma.

Lemma 1.7.1. *Set $\mu_{D,\varepsilon}(dx) = \Psi_{D,\varepsilon}(x)\mu(dx)$. If $\phi \in W_\mu^{m,p}(D)$, then $\phi \in W_{\mu_{D,\varepsilon}}^{m,p}(\mathbb{R}^d)$. Moreover, in the case $m = 1$, we get*

$$\partial_i^{\mu_{D,\varepsilon}}\phi = \partial_i^\mu\phi + \phi\partial_i\ln\Psi_{D,\varepsilon}, \quad i = 1,\ldots,d,$$

and

$$\|\phi\|_{W_{\mu_{D,\varepsilon}}^{1,p}(\mathbb{R}^d)} \leq C_p\varepsilon^{-1}\|\phi\|_{W_\mu^{1,p}(D)}.$$

Proof. If $f \in C_c^\infty(\mathbb{R}^d)$, then $f\Psi_{D,\varepsilon} \in C_c^\infty(D)$ and, similarly to what was developed in Lemma 1.1.1, we have

$$\int \partial_i f \phi \, d\mu_{D,\varepsilon} = \int \partial_i f \phi \Psi_{D,\varepsilon} d\mu = -\int \partial_i^\mu(\Psi_{D,\varepsilon}\phi)f\, d\mu$$

$$= -\int \left(\Psi_{D,\varepsilon}\partial_i^\mu\phi + \phi\partial_i\Psi_{D,\varepsilon}\right)f\, d\mu = -\int \left(\partial_i^\mu\phi + \phi\partial_i\ln\Psi_{D,\varepsilon}\right)f\, d\mu_{D,\varepsilon},$$

so that $\partial_i^{\mu_{D,\varepsilon}}\phi = \partial_i^\mu\phi + \phi\partial_i\ln\Psi_{D,\varepsilon}$. Now using (1.22) we have $\partial_i^{\mu_{D,\varepsilon}}\phi \in L_{\mu_{D,\varepsilon}}^p(\mathbb{R}^d)$:

$$\sum_{i=1}^d \int |\partial_i^{\mu_{D,\varepsilon}}\phi|^p d\mu_{D,\varepsilon} = \sum_{i=1}^d \int |\partial_i^\mu\phi + \phi\partial_i\ln\Psi_{D,\varepsilon}|^p\Psi_{D,\varepsilon}d\mu$$

$$\leq \sum_{i=1}^d \int_D |\partial_i^\mu\phi + \phi\partial_i\ln\Psi_{D,\varepsilon}|^p\Psi_{D,\varepsilon}d\mu \leq C_p\varepsilon^{-p}\|\phi\|_{W_\mu^{1,p}(D)}^p,$$

and the required estimate for $\|\phi\|_{W_{\mu_{D,\varepsilon}}^{1,p}(\mathbb{R}^d)}$ holds. $\square$

We are now ready to study the link between Sobolev spaces on open sets and "local" densities, and we collect all the results in the following theorem. The symbol $\nu|_D$ denotes the measure ν restricted to the open set D.

Theorem 1.7.2.   (i) *Suppose that $\phi \in W_\mu^{1,1}(D)$. Then $\mu_\phi|_D$ is absolutely continuous: $\mu_\phi|_D(dx) = p_{\phi,D}(x)dx$.*

(ii) *Suppose that $1 \in W_\mu^{1,p}(D)$ for some $p > d$. Then, for each $\varepsilon > 0$,*

$$\sup_{x\in\mathbb{R}^d}\sum_{i=1}^d \left(\int_{D_{2\varepsilon}} |\partial_i Q_d(y-x)|^{p/(p-1)}\mu(dy)\right)^{\frac{p-1}{p}} \leq C_{d,p}\varepsilon^{-k_{d,p}}\|1\|_{W_\mu^{1,p}(D)}^{k_{d,p}},$$

where $k_{d,p}$ is given in Theorem 1.4.1.

(iii) *Suppose that $1 \in W_\mu^{1,p}(D)$ for some $p > d$. Then, for $\phi \in W_\mu^{m,p}(D)$ we have $\mu_\phi|_D(dx) = p_{\phi,D}(x)dx$ and*

$$p_{\phi,D}(x) = -\sum_{i=1}^d \int \partial_i Q_d(y-x)(\Psi_{D,\varepsilon}\partial_i^\mu\phi + \phi\partial_i\Psi_{D,\varepsilon})\mu(dx),$$

for $x \in D_{2\varepsilon}$. *Moreover*, $p_{\phi,D} \in \bigcap_{\varepsilon > 0} W^{m,p}(D_\varepsilon)$ *and is* $m - 1$ *times differentiable on* D.

Proof. (i) Set $\mu_{\phi,D,\varepsilon}(dx) := \phi \mu_{D,\varepsilon}(dx) \equiv \phi \Psi_{D,\varepsilon}(x)\mu(dx)$. By Lemma 1.7.1, we know that $\phi \in W^{1,1}_{\mu_{D,\varepsilon}}(\mathbb{R}^d)$. So, we can use Theorem 1.5.2 to obtain $\mu_{\phi,D,\varepsilon}(dx) = p_{\phi,D,\varepsilon}(x)dx$, with $p_{\phi,D,\varepsilon} \in W^{1,1}(\mathbb{R}^d)$. Now, if $f \in C_c^\infty(D)$, then for some $\varepsilon > 0$ the support of f is included in $D_{2\varepsilon}$, so that $\int f \phi d\mu = \int f \phi d\mu_{D,\varepsilon} = \int f p_{\phi,D,\varepsilon}(x)dx$. It follows that $\mu_\phi|_D(dx) = p_{\phi,D}(x)dx$ and that, for $x \in D_{2\varepsilon}$, we have $p_{\phi,D}(x) = p_{\phi,D,\varepsilon}(x)dx$.

(ii) We have

$$\int_{D_{2\varepsilon}} |\partial_i Q_d(x - y)|^{p/(p-1)} \mu(dy) \leq \int |\partial_i Q_d(x - y)|^{p/(p-1)} \Psi_{D,\varepsilon}(y)\mu(dy)$$

$$= \int |\partial_i Q_d(x - y)|^{p/(p-1)} \mu_{D,\varepsilon}(dy),$$

so that

$$\sup_{x \in \mathbb{R}^d} \sum_{i=1}^d \left(\int_{D_{2\varepsilon}} |\partial_i Q_d(y - x)|^{p/(p-1)} \mu(dy) \right)^{\frac{p-1}{p}} \leq \Theta_p(\mu_{D,\varepsilon}).$$

Now, if $1 \in W^{1,p}_\mu(D)$ with $p > d$, then by Lemma 1.7.1, $1 \in W^{1,p}_{\mu_{D,\varepsilon}}(\mathbb{R}^d)$ with $p > d$ and Theorem 1.4.1 yields

$$\sup_{x \in \mathbb{R}^d} \sum_{i=1}^d \left(\int_{D_{2\varepsilon}} |\partial_i Q_d(y - x)|^{p/(p-1)} \mu(dy) \right)^{\frac{p-1}{p}} \leq dK_{d,p} \|1\|^{k_{d,p}}_{W^{1,p}_{\mu_{D,\varepsilon}}}$$

$$\leq C_{d,p} \left(\varepsilon^{-1} \|1\|_{W^{1,p}_\mu(D)} \right)^{k_{d,p}},$$

the latter estimate following from Lemma 1.7.1. Notice that we could state upper bounds for the density and its derivatives in a similar way.

(iii) All the statements follows from the already proved fact that, for $x \in D_{2\varepsilon}$, $p_{\phi,D}(x) = p_{\phi,D,\varepsilon}(x)$. By Lemma 1.7.1, $1 \in W^{1,p}_{\mu_{D,\varepsilon}}(\mathbb{R}^d)$ and $\phi \in W^{m,p}_{\mu_{D,\varepsilon}}(\mathbb{R}^d)$, so the result follows by applying Theorem 1.5.2. Note that this result can also be much more detailed by including estimates, e.g., for $\|p_{\phi,D}\|_{W^{m,p}(D_\varepsilon)}$. $\qquad \square$

1.8   Random variables

The generalized Sobolev spaces we have worked with up to now have a natural application to the context of random variables: we see now the link between the integration by parts underlying these spaces and the integration by parts type formulas for random variables. There are several ways to rewrite the results of the previous sections in terms of random variables, as we will see in the sequel. Here we propose just a first easy way of looking at things.

Let $(\Omega, \mathcal{F}, \mathbb{P})$ be a probability space, and denote by $\mathbb{E}$ the expectation under $\mathbb{P}$. Let F and G be two random variables taking values in $\mathbb{R}^d$ and $\mathbb{R}$, respectively.

Definition 1.8.1. *Given a multiindex α and a power $p \geq 1$, we say that the integration by parts formula $\mathrm{IBP}_{\alpha,p}(F, G)$ holds if there exists a random variable $H_\alpha(F; G) \in L^p$ such that*

$$\mathrm{IBP}_{\alpha,p}(F, G) \qquad \mathbb{E}\big(\partial_\alpha f(F) G\big) = \mathbb{E}\big(f(F) H_\alpha(F; G)\big), \quad \forall f \in C_c^{|\alpha|}(\mathbb{R}^d).$$

We set $W_F^{m,p}$ as the space of the random variables $G \in L^p$ such that $\mathrm{IBP}_{\alpha,p}(F, G)$ holds for every multiindex α with $|\alpha| \leq m$.

For $G \in W_F^{m,p}$ we set

$$\partial_\alpha^F G = \mathbb{E}\big(H_\alpha(F; G) \mid F\big), \quad 0 \leq |\alpha| \leq m,$$

with the understanding that $\partial_\alpha^F G = \mathbb{E}(G \mid F)$ when $|\alpha| = 0$, and we define the norm

$$\|G\|_{W_F^{m,p}} = \sum_{0 \leq |\alpha| \leq m} \big(\mathbb{E}(|\partial_\alpha^F G|^p)\big)^{1/p}.$$

It is easy to see that $(W_F^{m,p}, \|\cdot\|_{W_F^{m,p}})$ is a Banach space.

Remark 1.8.2. Notice that $\mathbb{E}\big(|\partial_\alpha^F G|^p\big) \leq \mathbb{E}(|H_\alpha(F; G)|^p)$, so that $\partial_\alpha^F G$ is the weight of minimal L^p norm which verifies $\mathrm{IBP}_{\alpha,p}(F; G)$. In particular,

$$\|G\|_{W_F^{m,p}} \leq \|G\|_{L^p} + \sum_{1 \leq |\alpha| \leq m} \|H_\alpha(F; G)\|_{L^p},$$

and this last quantity is the one which naturally appears in concrete computations.

The link between $W_F^{m,p}$ and the abstract Sobolev space $W_\mu^{m,p}$ is easy to find. In fact, let μ_F denote the law of F and take

$$\mu(dx) = \mu_F(dx) \quad \text{and} \quad \phi(x) = \phi_G(x) := \mathbb{E}(G \mid F = x).$$

By recalling that $\mathbb{E}(\partial_\alpha f(F) G) = \mathbb{E}(\partial_\alpha f(F) \mathbb{E}(G \mid F)) = \mathbb{E}(\partial_\alpha f(F) \phi_G(F))$ and similarly $\mathbb{E}(f(F) H_\alpha(F; G)) = \mathbb{E}(f(F) \mathbb{E}(H_\alpha(F; G) \mid F))$, a re-writing in integral form of the integration by parts formula $\mathrm{IBP}_{\alpha,p}(F; G)$ immediately gives

$$W_F^{m,p} = \big\{G \in L^p : \phi_G \in W_{\mu_F}^{m,p}\big\} \quad \text{and} \quad \partial_\alpha^F G = (-1)^{|\alpha|} \partial_\alpha^{\mu_F} \phi_G(F).$$

And of course, $\|G\|_{W_F^{m,p}} = \|\phi_G\|_{W_{\mu_F}^{m,p}}$.

So, concerning the density of F (equivalently, of μ_F), we can summarize the results of Section 1.1 as follows.

Theorem 1.8.3. (i) *Suppose that* $1 \in W_F^{1,p}$ *for some* $p > d$. *Then* $\mu_F(dx) = p_F(x)dx$ *and* $p_F \in C_b(\mathbb{R}^d)$. *Moreover,*

$$\Theta_F(p) := \sup_{a \in \mathbb{R}^d} \sum_{i=1}^{d} \left(\mathbb{E}\big(|\partial_i Q_d(F-a)|^{p/(p-1)}\big) \right)^{(p-1)/p} \leq K_{d,p} \|1\|_{W_F^{1,p}}^{k_{d,p}},$$

$$\|p_F\|_\infty \leq K_{d,p} \|1\|_{W_F^{1,p}}^{1+k_{d,p}},$$

and

$$p_F(x) = -\sum_{i=1}^{d} \mathbb{E}\big(\partial_i Q_d(F-x)\partial_i^F 1\big) = \sum_{i=1}^{d} \mathbb{E}\big(\partial_i Q_d(F-x)H_i(F,1)\big).$$

(ii) *For any positive* $a < 1/d - 1/p$, *we have*

$$p_F(x) \leq \Theta_F(\bar{p})\big(d + \|1\|_{W_F^{1,p}}\big) \mathbb{P}\big(F \in B_2(x)\big)^a,$$

where $\bar{p} = 1/(a+1/p)$.

Next we give the representation formula and the estimates for the conditional expectation ϕ_G. To this purpose, we set

$$\mu_{F,G}(f) := \mathbb{E}\big(f(F)G\big)$$

so that $\mu_F = \mu_{F,1}$. Notice that, in the notations of the previous sections,

$$\mu_{F,G}(dx) = \mu_\phi(dx) = \phi(x)\mu(dx),$$

with $\mu = \mu_F$ and $\phi = \phi_G$. So, the results may be re-written in this context as in Theorem 1.8.4 below. In particular, we get a formula allowing to write ϕ_G in terms of the representation formulas for the density $p_{F,G}$ and p_F, of $\mu_{F,G}$ and μ_F, respectively. And since ϕ_G stands for the conditional expectation of G given F, the representation we are going to write down assumes an important role in probability. Let us also recall that Malliavin and Thalmaier firstly developed such a formula in [36].

Theorem 1.8.4. *Suppose* $1 \in W_F^{1,p}$. *Let* $m \geq 1$ *and* $G \in W_F^{m,p}$.

(i) *We have* $\mu_{F,G}(dx) = p_{F,G}(x)dx$ *and*

$$\phi_G(x) = \mathbb{E}\big(G \mid F = x\big) = 1_{\{p_F(x)>0\}} \frac{p_{F,G}(x)}{p_F(x)},$$

with

$$p_{F,G}(x) = \sum_{i=1}^{d} \mathbb{E}\big(\partial_i Q_d(F-x)\partial_i^F G\big) = \sum_{i=1}^{d} \mathbb{E}\big(\partial_i Q_d(F-x)H_i(F;G)\big).$$

(ii) *We have $p_{F,G} \in W^{m,p}$ and $\phi_G \in \bigcap_{\delta>0} W^{m,p}(\{p_F > \delta\})$. Moreover,*

    (a) $\|p_{F,G}\|_\infty \leq K_{d,p} \|1\|_{W_F^{1,p}}^{k_{d,p}} \|G\|_{W_F^{1,p}},$

    (b) $\|p_{F,G}\|_{W^{m,p}} \leq (2dK_{d,p})^{1-1/p} \|1\|_{W_F^{1,p}}^{k_{d,p}(1-1/p)} \|G\|_{W_F^{m,p}}.$

We also have the representation formula

$$\partial_\alpha p_{F,G}(x) = \sum_{i=1}^d \mathbb{E}\big(\partial_i Q_d(F-x)\partial_{(\alpha,i)}^F G\big) = \sum_{i=1}^d \mathbb{E}\big(\partial_i Q_d(F-x)H_{(\alpha,i)}(F;G)\big)$$

for any α with $0 \leq |\alpha| \leq m-1$. Furthermore, $p_{F,G} \in C^{m-1}(\mathbb{R}^d)$ and for any multiindex α with $|\alpha| = k \leq m-2$, $\partial_\alpha p_{F,G}$ is Lipschitz continuous with Lipschitz constant

$$L_\alpha = d^2 \|1\|_{W_F^{1,p}}^{k_{d,p}} \|G\|_{W_F^{k+2,p}}.$$

Moreover, for any multiindex α with $|\alpha| = m-1$, $\partial_\alpha p_{F,G}$ is Hölder continuous of exponent $1 - d/p$ and Hölder constant

$$L_\alpha = C_{d,p} \|p_{F,G}\|_{W^{m,p}},$$

where $C_{d,p}$ depends only on d and p.

Finally we give a stability property.

Proposition 1.8.5. *Let F_n, G_n, $n \in \mathbb{N}$, be two sequences of random variables such that $(F_n, G_n) \to (F, G)$ in probability. Suppose that $G_n \in W_{F_n}^{m,p}$ and*

$$\sup_n (\|G_n\|_{W_{F_n}^{m,p}} + \|F_n\|_{L^p}) < \infty$$

for some $m \in \mathbb{N}$. Then, $G \in W_F^{m,p}$ and $\|G\|_{W_F^{m,p}} \leq \sup_n \|G_n\|_{W_{F_n}^{m,p}}.$

Remark 1.8.6. Proposition 1.8.5 may be used to get integration by parts formulas starting from the standard integration by parts formulas for the Lebesgue measure. In fact, let F_n, G_n, $n \in \mathbb{N}$, be a sequence of simple functionals (e.g., smooth functions of the noise) such that $F_n \to F$ and $G_n \to G$. By assuming the noise has a density, we can use standard finite-dimensional integration by parts formulas to prove that $G_n \in W_{F_n}^{m,p}$. If we are able to check that $\sup_n \|G_n\|_{W_{F_n}^{m,p}} < \infty$, then, using the above stability property, we can conclude that $G \in W_F^{m,p}$. Let us observe that the approximation of random variables by simple functionals is quite standard in the framework of Malliavin calculus (not necessarily on the Wiener space).

Proof of Proposition 1.8.5. Write $Q_n = (F_n, G_n, \partial_\alpha^{F_n} G_n, |\alpha| \leq m)$. Since $p \geq 2$ and $\sup_n(\|G_n\|_{W_{F_n}^{m,p}} + \|F_n\|_{L^p}) < \infty$, the sequence Q_n, $n \in \mathbb{N}$, is bounded in L^2 and consequently weakly relatively compact. Let Q be a limit point. Using Mazur's theorem we construct a sequence of convex combinations $\overline{Q}_n = \sum_{i=1}^{k_n} \lambda_i^n Q_{n+i}$, (with $\sum_{i=1}^{k_n} \lambda_i^n = 1$ and $\lambda_i^n \geq 0$) such that $\overline{Q}_n \to Q$ strongly in L^2. Passing to a subsequence, we may assume that the convergence holds almost surely as well. Since $(F_n, G_n) \to (F, G)$ in probability, it follows that $Q = (F, G, \theta_\alpha, |\alpha| \leq m)$. Using the integration by parts formulas $\text{IBP}_{\alpha,p}(F_n, G_n)$ and the almost sure convergence it is easy to see that $\text{IBP}_{\alpha,p}(F, G)$ holds with $H_\alpha(F; G) = \theta_\alpha$ so, $\theta_\alpha = \partial_\alpha^F G$. Moreover, using again the almost sure convergence and the convex combinations, it is readily checked that $\|G\|_{W_F^{m,p}} \leq \sup_n \|G\|_{W_{F_n}^{m,p}}$. In all the above arguments we have to use the Lebesgue dominated convergence theorem, so the almost sure convergence is not sufficient. But a straightforward truncation argument which we do not develop here permits to handle this difficulty. □

Chapter 2

Construction of integration by parts formulas

In this chapter we construct integration by parts formulas in an abstract framework based on a finite-dimensional random vector $V = (V_1, \ldots, V_J)$; we follow [8]. Such formulas have been used in [8, 10] in order to study the regularity of solutions of jump-type stochastic equations, that is, including equations with discontinuous coefficients for which the Malliavin calculus developed by Bismut [17] and by Bichteler, Gravereaux, and Jacod [16] fails.

But our interest in these formulas here is also "methodological": we want to understand better what is going on in Malliavin calculus. The framework is the following: we consider functionals of the form $F = f(V)$, where f is a smooth function. Our basic assumption is that the law of V is absolutely continuous with density $p_J \in C^\infty(\mathbb{R}^J)$. In Malliavin calculus V is Gaussian (it represents the increments of the Brownian path on a finite time grid) and the functionals F defined as before are the so-called "simple functionals". We mimic the Malliavin calculus in this abstract framework: so we define differential operators as there —but now we find out standard finite-dimensional gradients instead of the classical Malliavin derivative. The outstanding consequence of keeping the differential structure from Malliavin calculus is that we are able to construct objects for which we obtain estimates which are independent of the dimension J of the noise V. This will permit to recover the classical Malliavin calculus using this finite-dimensional calculus. Another important point is that the density p_J comes on in the calculus by means of $\partial \ln p_J(x)$ only —in the Gaussian case (one-dimensional for simplicity) this is $-x/\sigma^2$, where σ^2 is the variance of the Gaussian random variable at hand. This specific structure explains the construction of the divergence operator in Malliavin calculus and is also responsible of the fact that the objects fit well in an infinite-dimensional structure. However, for a general random variable V with density p_J such nice properties may fail and it is not clear how to pass to the infinite-dimensional case. Nevertheless these finite-dimensional integration by parts formulas remain interesting and may be useful in order to study the regularity of the laws of random variables —we do this in Section 2.2.

2.1   Construction of integration by parts formulas

In this section we give an elementary construction of integration by parts formulas
for functionals of a finite-dimensional noise which mimic the infinite-dimensional
Malliavin calculus (we follow [8]). We stress that the operators we introduce represent the finite-dimensional variant of the derivative and the divergence operators
from the classical Malliavin calculus —and as an outstanding consequence all the
constants which appear in the estimates do not depend on the dimension of the
noise.

2.1.1   Derivative operators

On a probability space $(\Omega, \mathcal{F}, P)$, fix a J-dimensional random vector representing
the basic noise, $V = (V_1, \ldots, V_J)$. Here, $J \in \mathbb{N}$ is a deterministic integer. For each
$i = 1, \ldots, J$ we consider two constants $-\infty \leq a_i < b_i \leq \infty$ and we point out that
they may be equal to $-\infty$, respectively to $+\infty$. We denote

$$O_i = \big\{ v = (v_1, \ldots, v_J) : a_i < v_i < b_i, \quad i = 1, \ldots, J \big\}.$$

Our basic hypothesis will be that the law of V is absolutely continuous with respect
to the Lebesgue measure on $\mathbb{R}^J$ and the density p_J is smooth with respect to v_i
on the set O_i. The natural example which comes on in the standard Malliavin
calculus is the Gaussian law on $\mathbb{R}^J$. In this case $a_i = -\infty$ and $b_i = +\infty$. But we
may also take (as an example) V_i independent random variables with exponential
law and, in this case, $p_J(v) = \prod_{i=1}^{J} e_i(v_i)$ with $e_i(v_i) = 1_{(0,\infty)}(v_i)e^{-v_i}$. Since e_i is
discontinuous in zero, we will take $a_i = 0$ and $b_i = \infty$.

In order to obtain integration by parts formulas for functionals of V, we
will perform classical integration by parts with respect to $p_J(v)dv$. But boundary
terms may appear in a_i and b_i. In order to cancel them we will consider some
"weights"

$$\pi_i \colon \mathbb{R}^J \longrightarrow [0,1], \quad i = 1, \ldots, J.$$

These are just deterministic measurable functions (in [8] the weights π_i are allowed
to be random, but here, for simplicity, we consider just deterministic ones). We
give now the precise statement of our hypothesis.

Hypothesis 2.1.1. The law of the vector $V = (V_1, \ldots, V_J)$ is absolutely continuous
with respect to the Lebesgue measure on $\mathbb{R}^J$ and we note p_J the density. We also
assume that

(i) $p_J(v) > 0$ for $v \in O_i$ and $v_i \mapsto p_J(v)$ is of class C^∞ on O_i;

(ii) for every $q \geq 1$, there exists a constant C_q such that

$$(1 + |v|^q)p_J(v) \leq C_q,$$

where $|v|$ stands for the Euclidean norm of the vector $v = (v_1, \ldots, v_J)$;

(iii) $v_i \mapsto \pi_i(v)$ is of class C^∞ on O_i and

    (a) $\lim_{v_i \downarrow a_i} \pi_i(v) p_J(v) = \lim_{v_i \uparrow b_i} \pi_i(v) p_J(v) = 0$,

    (b) $\pi_i(v) = 0$ for $v \notin O_i$,

    (c) $\pi_i \partial_{v_i} \ln p_J \in C_p^\infty(\mathbb{R}^J)$,

where $C_p^\infty(\mathbb{R}^J)$ is the space of functions which are infinitely differentiable and have polynomial growth (together with their derivatives).

Simple functionals. A random variable F is called a simple functional if there exists $f \in C_p^\infty(\mathbb{R}^J)$ such that $F = f(V)$. We denote through $\mathcal{S}$ the set of simple functionals.

Simple processes. A simple process is a J-dimensional random vector, say $U = (U_1, \ldots, U_J)$, such that $U_i \in \mathcal{S}$ for each $i = 1, \ldots, J$. We denote by $\mathcal{P}$ the space of simple processes. On $\mathcal{P}$ we define the scalar product

$$\langle U, \bar{U} \rangle_J = \sum_{i=1}^{J} U_i \bar{U}_i, \quad U, \bar{U} \in \mathcal{P}.$$

Note that $\langle U, \bar{U} \rangle_J \in \mathcal{S}$.

We can now define the derivative operator and state the integration by parts formula.

The derivative operator. We define $D \colon \mathcal{S} \to \mathcal{P}$ by

$$DF := (D_i F)_{i=1,\ldots,J} \in \mathcal{P}, \tag{2.1}$$

where $D_i F := \pi_i(V)\partial_i f(V)$. Clearly, D depends on π so a correct notation should be $D_i^\pi F := \pi_i(V)\partial_i f(V)$. Since here the weights π_i are fixed, we do not indicate them in the notation.

The Malliavin covariance matrix of F is defined by

$$\sigma^{k,k'}(F) = \langle DF^k, DF^{k'} \rangle_J = \sum_{j=1}^{J} D_j F^k D_j F^{k'}.$$

We denote

$$\Lambda(F) = \{\, \det \sigma(F) \neq 0 \,\} \quad \text{and} \quad \gamma(F)(\omega) = \sigma^{-1}(F)(\omega), \quad \omega \in \Lambda(F).$$

The divergence operator.   Let $U = (U_1, \ldots, U_J) \in \mathcal{P}$, so that $U_i \in \mathcal{S}$ and $U_i = u_i(V)$, for some $u_i \in C_p^\infty(\mathbb{R}^J)$, $i = 1, \ldots, J$. We define $\delta \colon \mathcal{P} \to \mathcal{S}$ by

$$\delta(U) = \sum_{i=1}^J \delta_i(U), \text{ with } \delta_i(U) := -\big(\partial_{v_i}(\pi_i u_i) + \pi_i u_i 1_{O_i} \partial_{v_i} \ln p_J\big)(V), \; i = 1, \ldots, J.$$
$$(2.2)$$

The divergence operator depends on π_i so, the precise notations should be $\delta_i^\pi(U)$ and $\delta^\pi(U)$. But, as in the case of the derivative operator, we do not mention this in the notation.

The Ornstein–Uhlenbeck operator.   We define $L \colon \mathcal{S} \to \mathcal{S}$ by

$$L(F) = \delta(DF). \tag{2.3}$$

2.1.2   Duality and integration by parts formulas

In our framework the duality between δ and D is given by the following proposition.

Proposition 2.1.2. *Assume Hypothesis* 2.1.1. *Then, for every* $F \in \mathcal{S}$ *and* $U \in \mathcal{P}$, *we have*
$$\mathbb{E}\big(\langle DF, U \rangle_J\big) = \mathbb{E}(F\delta(U)). \tag{2.4}$$

Proof. Let $F = f(V)$ and $U_i = u_i(V)$. By definition, we have $\mathbb{E}(\langle DF, U \rangle_J) = \sum_{i=1}^J \mathbb{E}(D_i F \times U_i)$ and, from Hypothesis 2.1.1,

$$\mathbb{E}\big(D_i F \times U_i\big) = \int_{\mathbb{R}^J} \partial_{v_i}(f)\pi_i \, u_i \, p_J(v_1, \ldots, v_J) dv_1 \cdots dv_J.$$

Recalling that $\{\pi_i > 0\} \subset O_i$, we obtain from Fubini's theorem

$$\mathbb{E}\big(D_i F \times U_i\big) = \int_{\mathbb{R}^{J-1}} \left(\int_{a_i}^{b_i} \partial_{v_i}(f)\pi_i \, u_i \, p_J(v_1, \ldots, v_J) dv_i \right) dv_1 \cdots dv_{i-1} dv_{i+1} \cdots dv_J.$$

The classical integration by parts formula yields

$$\int_{a_i}^{b_i} \partial_{v_i}(f)\pi_i \, u_i \, p_J(v_1, \ldots, v_J) dv_i = \big[f\pi_i u_i p_J\big]_{a_i}^{b_i} - \int_{a_i}^{b_i} f \partial_{v_i}(u_i \pi_i p_J) dv_i.$$

By Hypothesis 2.1.1, $[f\pi_i u_i p_J]_{a_i}^{b_i} = 0$ so we obtain

$$\int_{a_i}^{b_i} \partial_{v_i}(f)\pi_i \, u_i \, p_J(v_1, \ldots, v_J) dv_i = - \int_{a_i}^{b_i} f \partial_{v_i}(u_i \pi_i p_J) dv_i.$$

The result follows by observing that

$$\partial_{v_i}(u_i \pi_i p_J) = (\partial_{v_i}(u_i \pi_i) + u_i 1_{O_i} \pi_i \partial_{v_i}(\ln p_J))p_J. \qquad \square$$

We have the following straightforward computation rules.

Lemma 2.1.3. (i) *Let* $\phi \in C_p^\infty(\mathbb{R}^d)$ *and* $F = (F^1, \ldots, F^d) \in \mathcal{S}^d$. *Then,* $\phi(F) \in \mathcal{S}$
and

$$D\phi(F) = \sum_{r=1}^{d} \partial_r \phi(F) DF^r. \tag{2.5}$$

(ii) *If* $F \in \mathcal{S}$ *and* $U \in \mathcal{P}$, *then*

$$\delta(FU) = F\delta(U) - \langle DF, U \rangle_J. \tag{2.6}$$

(iii) *For* $\phi \in C_p^\infty(\mathbb{R}^d)$ *and* $F = (F^1, \ldots, F^d) \in \mathcal{S}^d$, *we have*

$$L\phi(F) = \sum_{r=1}^{d} \partial_r \phi(F) LF^r - \sum_{r,r'=1}^{d} \partial_{r,r'} \phi(F) \langle DF^r, DF^{r'} \rangle_J. \tag{2.7}$$

Proof. (i) is a consequence of the chain rule, whereas (ii) follows from the definition of the divergence operator δ (one may alternatively use the chain rule for the derivative operator and the duality formula). Combining these equalities we get (iii). □

We can now state the main results of this section.

Theorem 2.1.4. *We assume Hypothesis 2.1.1. Let* $F = (F^1, \ldots, F^d) \in \mathcal{S}^d$ *be such that* $\Lambda(F) = \Omega$ *and*

$$\mathbb{E}(|\det \gamma(F)|^p) < \infty \quad \forall p \geq 1. \tag{2.8}$$

Then, for every $G \in \mathcal{S}$ *and for every smooth function* $\phi: \mathbb{R}^d \to \mathbb{R}$ *with bounded derivatives,*

$$\mathbb{E}(\partial_r \phi(F) G) = \mathbb{E}(\phi(F) H_r(F, G)), \quad r = 1, \ldots, d, \tag{2.9}$$

with

$$H_r(F, G) = \sum_{r'=1}^{d} \delta(G\gamma^{r',r}(F) DF^{r'})$$

$$= \sum_{r'=1}^{d} \left(G\delta(\gamma^{r',r}(F) DF^{r'}) - \gamma^{r',r}(F) \langle DF^{r'}, DG \rangle_J \right). \tag{2.10}$$

Proof. Using the chain rule we get

$$\langle D\phi(F), DF^{r'} \rangle_J = \sum_{j=1}^{J} D_j \phi(F) D_j F^{r'}$$

$$= \sum_{j=1}^{J} \left(\sum_{r=1}^{d} \partial_r \phi(F) D_j F^r \right) D_j F^{r'} = \sum_{r=1}^{d} \partial_r \phi(F) \sigma^{r,r'}(F),$$

so that $\partial_r \phi(F) = \sum_{r'=1}^{d} \langle D\phi(F), DF^{r'} \rangle_J \gamma^{r',r}(F)$. Since $F \in \mathcal{S}^d$, it follows that $\phi(F) \in \mathcal{S}$ and $\sigma^{r,r'}(F) \in \mathcal{S}$. Moreover, since $\det \gamma(F) \in \bigcap_{p \geq 1} L^p$, it follows that $G\gamma^{r',r}(F)DF^{r'} \in \mathcal{P}$ and the duality formula gives

$$\mathbb{E}\big(\partial_r \phi(F)G\big) = \sum_{r'=1}^{d} \mathbb{E}\big(\langle D\phi(F), G\gamma^{r',r}(F)DF^{r'} \rangle_J\big)$$

$$= \sum_{r'=1}^{d} \mathbb{E}\big(\phi(F)\delta(G\gamma^{r',r}(F)DF^{r'})\big). \qquad \square$$

We can extend the above integration by parts formula to higher-order derivatives.

Theorem 2.1.5. *Under the assumptions of Theorem 2.1.4, for every multiindex* $\beta = (\beta_1, \ldots, \beta_q) \in \{1, \ldots, d\}^q$ *we have*

$$\mathbb{E}\big(\partial_\beta \phi(F)G\big) = \mathbb{E}\big(\phi(F)H_\beta^q(F,G)\big), \qquad (2.11)$$

where the weights $H_\beta^q(F,G)$ *are defined recursively by (2.10) and*

$$H_\beta^q(F,G) = H_{\beta_1}\big(F, H_{(\beta_2,\ldots,\beta_q)}^{q-1}(F,G)\big). \qquad (2.12)$$

Proof. The proof is straightforward by induction. For $q = 1$, this is just Theorem 2.1.4. Now assume that the statement is true for $q \geq 1$ and let us prove it for $q + 1$. Let $\beta = (\beta_1, \ldots, \beta_{q+1}) \in \{1, \ldots, d\}^{q+1}$. Then we have

$$\mathbb{E}\big(\partial_\beta \phi(F)G\big) = \mathbb{E}\big(\partial_{(\beta_2,\ldots,\beta_{q+1})}(\partial_{\beta_1}\phi(F))G\big) = \mathbb{E}\big(\partial_{\beta_1}\phi(F)H_{(\beta_2,\ldots,\beta_{q+1})}^q(F,G)\big),$$

and the result follows. $\qquad \square$

Conclusion. The calculus presented above represents a standard finite-dimensional differential calculus. Nevertheless, it will be clear in the following section that the differential operators are defined in such a way that "everything is independent of the dimension J of V". Moreover, we notice that the only way in which the law of V is involved in this calculus is by means of $\nabla \ln p_J$.

2.1.3   Estimation of the weights

Iterated derivative operators, Sobolev norms

In order to estimate the weights $H_\beta^q(F,G)$ appearing in the integration by parts formulas of Theorem 2.1.5, we need first to define iterations of the derivative operator. Let $\alpha = (\alpha_1, \ldots, \alpha_k)$ be a multiindex, with $\alpha_i \in \{1, \ldots, J\}$, for $i = 1, \ldots, k$, and $|\alpha| = k$. Let $F \in \mathcal{S}$. For $k = 1$ we set $D_\alpha^1 F = D_\alpha F$ and $D^1 F = (D_\alpha F)_{\alpha \in \{1,\ldots,J\}}$. For $k > 1$, we define recursively

$$D_{(\alpha_1,\ldots,\alpha_k)}^k F = D_{\alpha_k}\big(D_{(\alpha_1,\ldots,\alpha_{k-1})}^{k-1}F\big) \quad \text{and} \quad D^k F = \big(D_{(\alpha_1,\ldots,\alpha_k)}^k F\big)_{\alpha_i \in \{1,\ldots,J\}}.$$
$$(2.13)$$

Remark that $D^k F \in \mathbb{R}^{J \otimes k}$ and, consequently, we define the norm of $D^k F$ as

$$|D^k F| = \left(\sum_{\alpha_1, \ldots, \alpha_k = 1}^{J} |D^k_{(\alpha_1, \ldots, \alpha_k)} F|^2 \right)^{1/2}. \tag{2.14}$$

Moreover, we introduce the following norms for simple functionals: for $F \in \mathcal{S}$ we set

$$|F|_{1,l} = \sum_{k=1}^{l} |D^k F| \quad \text{and} \quad |F|_l = |F| + |F|_{1,l} = \sum_{k=0}^{l} |D^k F| \tag{2.15}$$

and, for $F = (F^1, \ldots, F^d) \in \mathcal{S}^d$,

$$|F|_{1,l} = \sum_{r=1}^{d} |F^r|_{1,l} \quad \text{and} \quad |F|_l = \sum_{r=1}^{d} |F^r|_l.$$

Similarly, for $F = (F^{r,r'})_{r,r'=1,\ldots,d}$ we set

$$|F|_{1,l} = \sum_{r,r'=1}^{d} |F^{r,r'}|_{1,l} \quad \text{and} \quad |F|_l = \sum_{r,r'=1}^{d} |F^{r,r'}|_l.$$

Finally, for $U = (U_1, \ldots, U_J) \in \mathcal{P}$, we set $D^k U = (D^k U_1, \ldots, D^k U_J)$ and we define the norm of $D^k U$ as

$$|D^k U| = \left(\sum_{i=1}^{J} |D^k U_i|^2 \right)^{1/2}.$$

We allow the case $k = 0$, giving $|U| = (\langle U, U \rangle_J)^{1/2}$. Similarly to (2.15), we set

$$|U|_{1,l} = \sum_{k=1}^{l} |D^k U| \quad \text{and} \quad |U|_l = |U| + |U|_{1,l} = \sum_{k=0}^{l} |D^k U|.$$

Observe that for $F, G \in \mathcal{S}$, we have $D(F \times G) = DF \times G + F \times DG$. This leads to the following useful inequalities.

Lemma 2.1.6. *Let $F, G \in \mathcal{S}$ and $U, \bar{U} \in \mathcal{P}$. We have*

$$|F \times G|_l \leq 2^l \sum_{l_1 + l_2 \leq l} |F|_{l_1} |G|_{l_2}, \tag{2.16}$$

$$|\langle U, \bar{U} \rangle_J|_l \leq 2^l \sum_{l_1 + l_2 \leq l} |U|_{l_1} |\bar{U}|_{l_2}. \tag{2.17}$$

Let us remark that the first inequality is sharper than the inequality $|F \times G|_l \leq C_l |F|_l |G|_l$. Moreover, from (2.17) with $U = DF$ and $V = DG$ ($F, G, \in \mathcal{S}$) we deduce

$$|\langle DF, DG \rangle_J|_l \leq 2^l \sum_{l_1 + l_2 \leq l} |F|_{1,l_1+1} |G|_{1,l_2+1} \tag{2.18}$$

and, as an immediate consequence of (2.16) and (2.18), we have

$$|H \langle DF, DG \rangle_J|_l \leq 2^{2l} \sum_{l_1+l_2+l_3 \leq l} |F|_{1,l_1+1} |G|_{1,l_2+1} |H|_{l_3}. \qquad (2.19)$$

for $F, G, H \in \mathcal{S}$.

Proof of Lemma 2.1.6. We just prove (2.17), since (2.16) can be proved in the same way. We first give a bound for $D^k \langle U, \bar{U} \rangle_J = (D_\alpha^k \langle U, \bar{U} \rangle_J)_{\alpha \in \{1,\dots,J\}^k}$. For a multiindex $\alpha = (\alpha_1, \dots, \alpha_k)$ with $\alpha_i \in \{1, \dots, J\}$, we note $\alpha(\Gamma) = (\alpha_i)_{i \in \Gamma}$, where $\Gamma \subset \{1, \dots, k\}$ and $\alpha(\Gamma^c) = (\alpha_i)_{i \notin \Gamma}$. We have

$$D_\alpha^k \langle U, \bar{U} \rangle_J = \sum_{i=1}^J D_\alpha^k (U_i \bar{U}_i) = \sum_{k'=0}^k \sum_{|\Gamma|=k'} \sum_{i=1}^J D_{\alpha(\Gamma)}^{k'} U_i \times D_{\alpha(\Gamma^c)}^{k-k'} \bar{U}_i.$$

Let $W^{i,\Gamma} = (W_\alpha^{i,\Gamma})_{\alpha \in \{1,\dots,J\}^k} = (D_{\alpha(\Gamma)}^{k'} U_i \times D_{\alpha(\Gamma^c)}^{k-k'} \bar{U}_i)_{\alpha \in \{1,\dots,J\}^k}$. Then we have the following equality in $\mathbb{R}^{J \otimes k}$:

$$D^k \langle U, \bar{U} \rangle_J = \sum_{k'=0}^k \sum_{|\Gamma|=k'} \sum_{i=1}^J W^{i,\Gamma}.$$

This gives

$$\left| D^k \langle U, \bar{U} \rangle_J \right| \leq \sum_{k'=0}^k \sum_{|\Gamma|=k'} \left| \sum_{i=1}^J W^{i,\Gamma} \right|,$$

where

$$\left| \sum_{i=1}^J W^{i,\Gamma} \right|^2 = \sum_{\alpha_1,\dots,\alpha_k=1}^J \left| \sum_{i=1}^J W_\alpha^{i,\Gamma} \right|^2.$$

By the Cauchy–Schwarz inequality,

$$\left| \sum_{i=1}^J W_\alpha^{i,\Gamma} \right|^2 = \left| \sum_{i=1}^J D_{\alpha(\Gamma)}^{k'} U_i \times D_{\alpha(\Gamma^c)}^{k-k'} \bar{U}_i \right|^2 \leq \sum_{i=1}^J \left| D_{\alpha(\Gamma)}^{k'} U_i \right|^2 \times \sum_{i=1}^J \left| D_{\alpha(\Gamma^c)}^{k-k'} \bar{U}_i \right|^2.$$

Consequently, we obtain

$$\left| \sum_{i=1}^J W^{i,\Gamma} \right|^2 \leq \sum_{\alpha_1,\dots,\alpha_k=1}^J \sum_{i=1}^J |D_{\alpha(\Gamma)}^{k'} U_i|^2 \times \sum_{i=1}^J |D_{\alpha(\Gamma^c)}^{k-k'} \bar{U}_i|^2$$

$$= \left| D^{k'} U \right|^2 \times \left| D^{k-k'} \bar{U} \right|^2.$$

This last equality follows from the fact that we sum over different index sets Γ and Γ^c. This gives

$$
\big|D^k \langle U, \bar{U} \rangle_J\big| \leq \sum_{k'=0}^{k} \sum_{|\Gamma|=k'} |D^{k'}U|\,|D^{k-k'}\bar{U}| = \sum_{k'=0}^{k} C_k^{k'} |D^{k'}U|\,|D^{k-k'}\bar{U}|
$$

$$
\leq \sum_{k'=0}^{k} C_k^{k'} |U|_{k'} |\bar{U}|_{k-k'} \leq 2^k \sum_{l_1+l_2=k} |U|_{l_1} |\bar{U}|_{l_2}.
$$

Summing over $k = 0, \dots, l$ we deduce (2.18). $\qquad\qquad\square$

Estimate of $|\gamma(F)|_l$

We give in this subsubsection an estimation of the derivatives of $\gamma(F)$ in terms of $\det \sigma_F$ and the derivatives of F. We assume that $\omega \in \Lambda(F)$.

In what follows $C_{l,d}$ is a constant, possibly depending on the order of derivation l and the dimension d, but not on J.

For $F \in \mathcal{S}^d$, we set

$$
m(\sigma_F) = \max\left(1, \frac{1}{\det \sigma_F}\right). \tag{2.20}
$$

Proposition 2.1.7. (i) *If $F \in \mathcal{S}^d$, then $\forall l \in \mathbb{N}$ we have*

$$
|\gamma(F)|_l \leq C_{l,d}\, m(\sigma_F)^{l+1} \big(1 + |F|_{1,l+1}^{2d(l+1)}\big). \tag{2.21}
$$

(ii) *If $F, \overline{F} \in \mathcal{S}^d$, then $\forall l \in \mathbb{N}$ we have*

$$
|\gamma(F) - \gamma(\overline{F})|_l
$$

$$
\leq C_{l,d}\, m(\sigma_F)^{l+1}\, m(\sigma_{\overline{F}})^{l+1} \big(1 + |F|_{1,l+1} + |\overline{F}|_{1,l+1}\big)^{2d(l+3)} \big|F - \overline{F}\big|_{1,l+1}. \tag{2.22}
$$

Before proving Proposition 2.1.7, we establish a preliminary lemma.

Lemma 2.1.8. *For every $G \in S$, $G > 0$, we have*

$$
\left|\frac{1}{G}\right|_l \leq C_l\left(\frac{1}{G} + \sum_{k=1}^{l} \frac{1}{G^{k+1}} \sum_{\substack{k \leq r_1+\cdots+r_k \leq l \\ r_1,\dots,r_k \geq 1}} \prod_{i=1}^{k} |D^{r_i}G|\right) \leq C_l\left(\frac{1}{G} + \sum_{k=1}^{l} \frac{1}{G^{k+1}} |G|_{1,l}^k\right).
$$

$$
\tag{2.23}
$$

Proof. For $F \in \mathcal{S}^d$ and $\phi \colon \mathbb{R}^d \to \mathbb{R}$ a $\mathcal{C}^\infty$ function, we have from the chain rule

$$
D^k_{(\alpha_1,\dots,\alpha_k)}\phi(F) = \sum_{|\beta|=1}^{k} \partial_\beta \phi(F) \sum_{\Gamma_1 \cup \cdots \cup \Gamma_{|\beta|} = \{1,\dots,k\}} \left(\prod_{i=1}^{|\beta|} D^{|\Gamma_i|}_{\alpha(\Gamma_i)} F^{\beta_i}\right), \tag{2.24}
$$

where $\beta \in \{1,\ldots,d\}^{|\beta|}$ and $\sum_{\Gamma_1 \cup \cdots \cup \Gamma_{|\beta|}}$ denotes the sum over all partitions of $\{1,\ldots,k\}$ with length $|\beta|$. In particular, for $G \in \mathcal{S}$, $G > 0$, and for $\phi(x) = 1/x$, we obtain

$$\left| D_\alpha^k \left(\frac{1}{G} \right) \right| \leq C_k \sum_{k'=1}^{k} \frac{1}{G^{k'+1}} \sum_{\Gamma_1 \cup \cdots \cup \Gamma_{k'}=\{1,\ldots,k\}} \left(\prod_{i=1}^{k'} |D_{\alpha(\Gamma_i)}^{|\Gamma_i|} G| \right). \tag{2.25}$$

We then deduce that

$$\left| D^k \left(\frac{1}{G} \right) \right| \leq C_k \sum_{k'=1}^{k} \frac{1}{G^{k'+1}} \sum_{\Gamma_1 \cup \cdots \cup \Gamma_{k'}=\{1,\ldots,k\}} \left| \prod_{i=1}^{k'} D_{\alpha(\Gamma_i)}^{|\Gamma_i|} G \right|_{\mathbb{R}^{J \otimes k}}$$

$$= C_k \sum_{k'=1}^{k} \frac{1}{G^{k'+1}} \sum_{\Gamma_1 \cup \cdots \cup \Gamma_{k'}=\{1,\ldots,k\}} \left(\prod_{i=1}^{k'} |D^{|\Gamma_i|} G| \right)$$

$$= C_k \sum_{k'=1}^{k} \frac{1}{G^{k'+1}} \sum_{\substack{r_1+\cdots+r_{k'}=k \\ r_1,\ldots,r_{k'} \geq 1}} \left(\prod_{i=1}^{k'} |D^{r_i} G| \right),$$

and the first part of (2.23) is proved. The proof of the second part is straightforward. $\qquad\square$

Proof of Proposition 2.1.7. (i) On $\Lambda(F)$ we have that

$$\gamma^{r,r'}(F) = \frac{1}{\det \sigma(F)} \hat{\sigma}^{r,r'}(F),$$

where $\hat{\sigma}(F)$ is the algebraic complement of $\sigma(F)$. But, recalling that $\sigma^{r,r'}(F) = \langle D^r F, D^{r'} F \rangle_J$, we have

$$\left| \det \sigma(F) \right|_l \leq C_{l,d} |F|_{1,l+1}^{2d} \quad \text{and} \quad |\hat{\sigma}(F)|_l \leq C_{l,d} |F|_{1,l+1}^{2(d-1)}. \tag{2.26}$$

Applying inequality (2.16), this gives

$$\left| \gamma(F) \right|_l \leq C_{l,d} \sum_{l_1+l_2 \leq l} \left| \left(\det \sigma(F) \right)^{-1} \right|_{l_1} |\hat{\sigma}(F)|_{l_2}.$$

From Lemma 2.1.8 and (2.26), it follows that

$$\left| \left(\det \sigma(F) \right)^{-1} \right|_{l_1} \leq C_{l_1} \left(\frac{1}{|\det \sigma(F)|} + \sum_{k=1}^{l_1} \frac{|F|_{1,l_1+1}^{2kd}}{|\det \sigma(F)|^{k+1}} \right)$$

$$\leq C_{l_1} m(\sigma_F)^{l_1+1} \left(1 + |F|_{1,l_1+1}^{2l_1 d} \right), \tag{2.27}$$

where we have used the constant $m(\sigma_F)$ defined in (2.20). Combining these inequalities, we obtain (2.21).

(ii) Using (2.18), we have

$$\left|\sigma^{r,r'}(F) - \sigma^{r,r'}(\overline{F})\right|_l \le C_{l,d}\left|F - \overline{F}\right|_{1,l+1}\left(|F|_{1,l+1} + |\overline{F}|_{1,l+1}\right)$$

and then, by (2.16),

$$\left|\det \sigma(F) - \det \sigma(\overline{F})\right|_l \le C_{l,d}\left|F - \overline{F}\right|_{1,l+1}\left(|F|_{1,l+1} + |\overline{F}|_{1,l+1}\right)^{2d-1} \qquad (2.28)$$

and

$$\left|\widehat{\sigma}^{r,r'}(F) - \widehat{\sigma}^{r,r'}(\overline{F})\right|_l \le C_{l,d}\left|F - \overline{F}\right|_{1,l+1}\left(|F|_{1,l+1} + |\overline{F}|_{1,l+1}\right)^{2d-3}.$$

Then, by using also (2.27), we have

$$\left|(\det \sigma(F))^{-1} - (\det \sigma(\overline{F}))^{-1}\right|_l$$

$$\le C_{l,d}\left|(\det \sigma(F))^{-1}\right|_l \left|(\det \sigma(\overline{F}))^{-1}\right|_l \times \times \left|\det \sigma(F) - \det \sigma(\overline{F})\right|_l$$

$$\le C_{l,d}m(\sigma_F)^{l+1}m(\sigma_{\overline{F}})^{l+1}|F - \overline{F}|_{1,l+1}\left(1 + |F|_{1,l} + |\overline{F}|_{1,l}\right)^{2(l+1)d}.$$

Since $\gamma^{r,r'}(F) = (\det \sigma(F))^{-1}\widehat{\sigma}^{r,r'}(F)$, a straightforward use of the above estimates gives

$$\left|\gamma^{r,r'}(F) - \gamma^{r,r'}(\overline{F})\right|_l$$

$$\le C_{l,d}m(\sigma_F)^{l+1}m(\sigma_{\overline{F}})^{l+1}|F - \overline{F}|_{1,l+1}\left(1 + |F|_{1,l+1} + |\overline{F}|_{1,l+1}\right)^{2(l+3)d}. \qquad \square$$

We define now

$$L_r^\gamma(F) = \sum_{r'=1}^d \delta\left(\gamma^{r',r}(F)DF^{r'}\right) = \sum_{r'=1}^d \left(\gamma^{r',r}(F)LF^{r'} - \langle D\gamma^{r',r}(F), DF^{r'}\rangle\right). \qquad (2.29)$$

Using (2.22), it can easily be checked that $\forall l \in \mathbb{N}$

$$|L_r^\gamma(F)|_l \le C_{l,d}m(\sigma_F)^{l+2}\left(1 + |F|_{l+1}^{2d(l+2)} + |L(F)|_l\right). \qquad (2.30)$$

And by using both (2.22) and (2.29), we immediately get

$$\left|L_r^\gamma(F) - L_r^\gamma(\overline{F})\right|_l \le C_{l,d}Q_l(F,\overline{F})\left(|F - \overline{F}|_{l+2} + |L(F - \overline{F})|_l\right), \qquad (2.31)$$

where

$$Q_l(F,\overline{F}) = m(\sigma_F)^{l+2}m(\sigma_{\overline{F}})^{l+2}\left(1 + |F|_{l+2}^{2d(l+4)} + |L(F)|_l + |\overline{F}|_{l+2}^{2d(l+4)} + |L(\overline{F})|_l\right). \qquad (2.32)$$

For $F \in \mathcal{S}^d$, we define the linear operator $T_r(F, \circ) \colon \mathcal{S} \to \mathcal{S}, r = 1, \ldots, d$ by

$$T_r(F, G) = \langle DG, (\gamma(F)DF)^r \rangle,$$

where $(\gamma(F)DF)^r = \sum_{r'=1}^d \gamma^{r',r}(F)DF^{r'}$. Moreover, for any multiindex $\beta = (\beta_1, \ldots, \beta_q)$, we denote $|\beta| = q$ and we define by induction

$$T_\beta(F, G) = T_{\beta_q}\big(F, T_{(\beta_1, \ldots, \beta_{q-1})}(F, G)\big).$$

For $l \in \mathbb{N}$ and $F, \overline{F} \in \mathcal{S}^d$, we denote

$$\Theta_l(F) = m(\sigma_F)^l \big(1 + |F|_{l+1}^{2d(l+1)}\big), \qquad\qquad\qquad (2.33)$$

$$\Theta_l(F, \overline{F}) = m(\sigma_F)^l m(\sigma_{\overline{F}})^l \big(1 + |F|_l^{2d(l+2)} + |\overline{F}|_l^{2d(l+2)}\big). \qquad (2.34)$$

We notice that $\Theta_l(F) \le \Theta_l(F, \overline{F})$, $\Theta_l(F) \le \Theta_{l+1}(F)$, and $\Theta_l(F, \overline{F}) \le \Theta_{l+1}(F, \overline{F})$.

Proposition 2.1.9. *Let $F, \overline{F}, G, \overline{G} \in \mathcal{S}^d$. Then, for every $l \in \mathbb{N}$ and for every multiindex β with $|\beta| = q \ge 1$, we have*

$$\big|T_\beta(F, G)\big|_l \le C_{l,d,q} \Theta_{l+q}^q(F)\, |G|_{l+q} \qquad\qquad\qquad (2.35)$$

and

$$\big|T_\beta(F, G) - T_\beta(\overline{F}, \overline{G})\big|_l$$
$$\le C_{l,d,q} \Theta_{l+q}^{\frac{q(q+1)}{2}}(F, \overline{F})\big(1 + |G|_{l+q} + |\overline{G}|_{l+q}\big)^q \times \big(|F - \overline{F}|_{l+q} + |G - \overline{G}|_{l+q}\big), \qquad\qquad (2.36)$$

where $C_{l,d,q}$ denotes a suitable constant depending on l, d, q and universal with respect to the involved random variables.

Proof. We start by proving (2.35). Hereafter, C denotes a constant, possibly varying and independent of the random variables which are involved. For $q = 1$ we have

$$|T_r(F, G)|_l \le C|DG|_l |\gamma(F)|_l |DF|_l \le C|\gamma(F)|_l |DF|_{l+1} |G|_{l+1}$$

and, by using (2.21), we get

$$\big|T_r(F, G)\big|_l \le Cm(\sigma_F)^{l+1}\big(1 + |F|_{l+1}^{2d(l+1)}\big)|DF|_{l+1} |G|_{l+1} \le C\Theta_{l+1}(F)|G|_{l+1}.$$

The proof for general β's follows by recurrence. In fact, for $|\beta| = q > 1$, we write $\beta = (\beta_{-q}, \beta_q)$ and we have

$$\big|T_\beta(F, G)\big|_l = \big|T_{\beta_q}(F, T_{\beta_{-q}}(F, G)\big|_l \le C\Theta_{l+1}(F)\big|T_{\beta_{-q}}(F, G)\big|_{l+1}$$
$$\le C\Theta_{l+1}(F)\Theta_{l+q}(F)^{q-1}|G|_{l+q} \le C\Theta_{l+q}(F)^q |G|_{l+q}.$$

We prove now (2.36), again by recurrence. For $|\beta| = 1$ we have

$$\left|T_r(F,G) - T_r(\overline{F},\overline{G})\right|_l \le \left|G - \overline{G}\right|_{l+1}\left(|F|_{l+1} + |\overline{F}|_{l+1}\right)\left(|\gamma(F)|_l + |\gamma(\overline{F})|_l\right)$$
$$+ \left|F - \overline{F}\right|_{l+1}\left(|G|_{l+1} + |\overline{G}|_{l+1}\right)\left(|\gamma(F)|_l + |\gamma(\overline{F})|_l\right)$$
$$+ \left|\gamma(F) - \gamma(\overline{F})\right|_l\left(|F|_{l+1} + |\overline{F}|_{l+1}\right)\left(|G|_{l+1} + |\overline{G}|_{l+1}\right).$$

Using (2.22), the last term can be bounded from above by

$$C\Theta_{l+1}(F,\overline{F})\left(|G|_{l+1} + |\overline{G}|_{l+1}\right)|F - \overline{F}|_{l+1}.$$

And, using again (2.22), the first two quantities in the right-hand side can be bounded, respectively, by

$$C\Theta_{l+1}(F,\overline{F})\left|G - \overline{G}\right|_{l+1}$$

and

$$C\Theta_{l+1}(F,\overline{F})\left(|G|_{l+1} + |\overline{G}|_{l+1}\right)|F - \overline{F}|_{l+1}.$$

So, we get

$$\left|T_r(F,G) - T_r(\overline{F},\overline{G})\right|_l \le C\Theta_{l+1}(F,\overline{F})\left(1 + |G|_{l+1} + |\overline{G}|_{l+1}\right)\left(|F - \overline{F}|_{l+1} + |G - \overline{G}|_{l+1}\right).$$

Now, for $|\beta| = q \ge 2$, we proceed by induction. In fact, we write

$$\left|T_\beta(F,G) - T_\beta(\overline{F},\overline{G})\right|_l =$$
$$= \left|T_{\beta_q}(F, T_{\beta-q}(F,G)) - T_{\beta_q}(F, T_{\beta-q}(\overline{F},\overline{G}))\right|_l$$
$$\le C\Theta_{l+1}(F,\overline{F})\left(1 + |T_{\beta-q}(F,G)|_{l+1} + |T_{\beta-q}(\overline{F},\overline{G})|_{l+1}\right)$$
$$\times \left(|F - \overline{F}|_{l+1} + |T_{\beta-q}(F,G) - T_{\beta-q}(\overline{F},\overline{G})|_{l+1}\right).$$

Now, by (2.35), we have

$$\left|T_{\beta-q}(F,G)\right|_{l+1} \le C\Theta_{l+q}^{q-1}(F)\,|G|_{l+q} \quad \text{and} \quad \left|T_{\beta-q}(\overline{F},\overline{G})\right|_{l+q} \le \Theta_{l+q}^{q-1}(\overline{F})\,|\overline{G}|_{l+q}.$$

Notice that

$$\Theta_{l+1}(F,\overline{F})\left(\Theta_{l+q}^{q-1}(F) + \Theta_{l+q}^{q-1}(\overline{F})\right) \le 2\Theta_{l+q}(F,\overline{F})^q$$

so, the factor in the first line of the last right-hand side can be bounded from above by

$$C\Theta_{l+q}(F,\overline{F})^q\left(1 + |G|_{l+q} + |\overline{G}|_{l+q}\right).$$

Moreover, by induction we have

$$\left|T_{\beta-q}(F,G) - T_{\beta-q}(\overline{F},\overline{G})\right|_{l+1} \le \Theta_{l+q}^{\frac{(q-1)q}{2}}(F,\overline{F})\left(1 + |G|_{l+q} + |\overline{G}|_{l+q}\right)^{q-1}$$
$$\times \left(|F - \overline{F}|_{l+q} + |G - \overline{G}|_{l+q}\right).$$

Finally, combining everything we get (2.36). $\qquad\square$

Bounds for the weights $H_\beta^q(F, G)$

Now our goal is to establish some estimates for the weights H^q in terms of the derivatives of G, F, LF, and $\gamma(F)$.

For $l \geq 1$ and $F, \overline{F} \in \mathcal{S}^d$, we set (with the understanding that $|\cdot|_0 = |\cdot|$)

$$A_l(F) = m(\sigma_F)^{l+1}\big(1 + |F|_{l+1}^{2d(l+2)} + |LF|_{l-1}\big), \tag{2.37}$$

$$A_l(F, \overline{F}) = m(\sigma_F)^{l+1} m(\sigma_{\overline{F}})^{l+1}\big(1 + |F|_{l+1}^{2d(l+3)} + |\overline{F}|_{l+1}^{2d(l+3)} + |LF|_{l-1} + |L\overline{F}|_{l-1}\big). \tag{2.38}$$

Theorem 2.1.10.   (i) *For $F \in \mathcal{S}^d$, $G \in \mathcal{S}$, $l \in \mathbb{N}$, and $q \in \mathbb{N}^*$, there exists a universal constant $C_{l,q,d}$ such that, for every multiindex $\beta = (\beta_1, \ldots, \beta_q)$,*

$$\left|H_\beta^q(F, G)\right|_l \leq C_{l,q,d} A_{l+q}(F)^q |G|_{l+q}, \tag{2.39}$$

where $A_{l+q}(F)$ is defined in (2.37).

(ii) *For $F, \overline{F} \in \mathcal{S}^d$, $G, \overline{G} \in \mathcal{S}$, and for all $q \in \mathbb{N}^*$, there exists an universal constant $C_{l,q,d}$ such that, for every multiindex $\beta = (\beta_1, \ldots, \beta_q)$,*

$$\left|H_\beta^q(F, G) - H_\beta^q(\overline{F}, \overline{G})\right|_l \leq C_{l,q,d} A_{l+q}(F, \overline{F})^{\frac{q(q+1)}{2}} \times \big(1 + |G|_{l+q} + |\overline{G}|_{l+q}\big)^q$$
$$\times \big(|F - \overline{F}|_{l+q+1} + |L(F - \overline{F})|_{l+q-1} + |G - \overline{G}|_{l+q}\big). \tag{2.40}$$

Proof. Let us first recall that

$$H_r(F, G) = GL_r^\gamma(F) - T_r(F, G),$$

$$H_\beta^q(F, G) = H_{\beta_1}(F, H_{(\beta_2, \ldots, \beta_q)}^{q-1}(F, G)).$$

Again, we let C denote a positive constant, possibly varying from line to line, but independent of the random variables $F, \overline{F}, G, \overline{G}$.

(i) Suppose $q = 1$ and $\beta = r$. Then,

$$|H_r(F, G)|_l \leq C|G|_l |L_r^\gamma(F)|_l + C|T_r(F, G)|_l.$$

By using (2.30) and (2.35), we can write

$$|H_r(F, G)|_l \leq Cm(\sigma_F)^{l+2}\big(1 + |F|_{l+1}^{2d(l+2)} + |LF|_l\big)|G|_l$$
$$+ m(\sigma_F)^{l+1}\big(1 + |F|_{l+2}^{2d(l+3)}\big)|G|_{l+1}$$
$$\leq A_{l+1}(F)|G|_{l+1}.$$

So, the statement holds for $q = 1$. And for $q > 1$ it follows by iteration:

$$|H_\beta^q(F, G)|_l = |H_{\beta_q}(F, H_{\beta_{-q}}^{q-1}(F, G))|_l \leq CA_{l+1}(F)|H_{\beta_{-q}}^{q-1}(F, G))|_{l+1}$$
$$\leq CA_{l+1}(F)A_{l+q}(F)^{q-1}|G|_{l+q};$$

since $A_{l+1}(F) \le A_{l+q}(F)$, the statement is proved.

(ii) We give a sketch of the proof for $q = 1$ only, the general case following by induction. Setting $\beta = r$, we have

$$\left| H_r(F, G) - H_r(\overline{F}, \overline{G}) \right|_l \le C \left(|L_r^\gamma(F)|_l |G - \overline{G}|_l + |\overline{G}|_l |L_r^\gamma(F) - L_r^\gamma(\overline{F})|_l \right.$$
$$\left. + |T_r(F, G) - T_r(\overline{F}, \overline{G})|_l \right).$$

Now, estimate (2.40) follows by using (2.30), (2.31), and (2.36). In the iteration for $q > 1$, it suffices to observe that $A_l(F) \le A_l(F, \overline{F})$, $A_l(F) \le A_{l+1}(F)$ and $A_l(F, \overline{F}) \le A_{l+1}(F, \overline{F})$. □

Conclusion. We stress once again that the constants appearing in the previous estimates are independent of J. Moreover, the estimates that are usually obtained/used in the standard Malliavin calculus concern L^p norms (i.e., with respect to the expectation $\mathbb{E}$). But here the estimates are even deterministic so, "pathwise".

2.1.4 Norms and weights

The derivative operator D and the divergence operator δ introduced in the previous sections depend on the weights π_i. As a consequence, the Sobolev norms $|\circ|_l$ depend also on π_i because they involve derivatives. And the dependence is rather intricate because, for example, in the computation of the second-order derivatives, derivatives of π_i are involved as well. The aim of this subsection is to replace these intricate norms with simpler ones. This is a step towards the L^2 norms used in the standard Malliavin calculus.

Up to now, we have not stressed the dependence on the weights π_i. We do this now. So, we recall that

$$D_i^\pi F := \pi_i(V) \partial_i f(V), \quad \delta_i^\pi(U) := -\left(\partial_{v_i}(\pi_i u_i) + \pi_i u_i 1_{O_i} \partial_{v_i} \ln p_J \right)(V),$$

and

$$D^\pi F = (D_i^\pi F)_{i=1,\ldots,J}, \quad \delta^\pi(U) = \sum_{i=1}^J \delta_i^\pi(U).$$

Now the notation D and δ is reserved for the case $\pi_i = 1$, so that

$$D_i F := \partial_i f(V) \quad \delta_i(U) := -\left(\partial_{v_i} u_i + u_i 1_{O_i} \partial_{v_i} \ln p_J \right)(V).$$

Let us specify the link between D^π and D, and between δ^π and δ. We define $T_\pi : \mathcal{P} \to \mathcal{P}$ by

$$T_\pi U = (\pi_i(V) U_i)_{i=1,\ldots,J} \quad \text{for} \quad U = (U_i)_{i=1,\ldots,J}.$$

Then

$$D^\pi F = T_\pi D F \quad \text{and} \quad \delta^\pi(U) = \delta(T_\pi U).$$

It follows that $L^\pi F = \delta^\pi(D^\pi F) = \delta(T_\pi T_\pi DF) = \delta(T_{\pi^2} DF)$ with $(\pi^2)_i = \pi_i^2$.

We go now further to higher-order derivatives. We have defined

$$D^{\pi,k}_{(\alpha_1,\ldots,\alpha_k)} F = D^\pi_{\alpha_k}\big(D^{\pi,k-1}_{(\alpha_1,\ldots,\alpha_{k-1})} F\big) \quad \text{and} \quad D^{\pi,k} F = \big(D^{\pi,k}_{(\alpha_1,\ldots,\alpha_k)} F\big)_{\alpha_i \in \{1,\ldots,J\}}.$$

Notice that

$$D^{\pi,2}_{(\alpha_1,\alpha_2)} F = D^\pi_{\alpha_2}\big(D^\pi_{\alpha_1} F\big) = (T_\pi D)_{\alpha_2}\big((T_\pi D)_{\alpha_1} F\big) = \pi_{\alpha_2}(V) D_{\alpha_2}\big(\pi_{\alpha_1}(V) D_{\alpha_1} F\big)$$

$$= \pi_{\alpha_2}(V)(D_{\alpha_2}\pi_{\alpha_1})(V) D_{\alpha_1} F + \pi_{\alpha_2}(V)\pi_{\alpha_1}(V) D_{\alpha_2} D_{\alpha_1} F.$$

This example shows that $D^{\pi,k} F$ has an intricate expression depending on π_i and their derivatives up to order $k-1$, and on the standard derivatives of F of order $j = 1,\ldots,k$. Of course, if π_i are just constants their derivatives are null and the situation is much simpler.

Let us now look to the Sobolev norms. We recall that in (2.15) we have defined $|F|_{1,\pi,l}$ and $|F|_{\pi,l}$ in the following way:

$$|D^{\pi,k} F| = \bigg(\sum_{\alpha_1,\ldots,\alpha_k=1}^{J} |D^{\pi,k}_{(\alpha_1,\ldots,\alpha_k)} F|^2 \bigg)^{1/2}$$

and

$$|F|_{1,\pi,l} = \sum_{k=1}^{l} |D^{\pi,k} F|, \quad |F|_{\pi,l} = |F| + |F|_{1,\pi,l} = \sum_{k=0}^{l} |D^{\pi,k} F|.$$

We introduce now the new norms by

$$\langle U, \bar{U} \rangle_\pi = \sum_{j=1}^{J} \pi_i^2 U_i \bar{U}_i = \langle T_\pi U, T_\pi \bar{U} \rangle_J, \quad |U|_\pi = \sqrt{\langle U, U \rangle_{\pi,J}}.$$

The relation with the norms considered before is given by

$$|DF|_\pi = |D^{\pi,1} F|.$$

But if we go further to higher-order derivatives this equality fails. Let us be more precise. Let

$$\mathcal{P}^{\otimes k} = \{U = (U_\alpha)_{|\alpha|=k} : U_\alpha = u_\alpha(V)\}.$$

Then $D^k \in \mathcal{P}^{\otimes k}$. On $\mathcal{P}^{\otimes k}$ we define

$$\langle U, \bar{U} \rangle_{\pi,k} = \sum_{|\alpha|=k} \bigg(\prod_{i=1}^{k} \pi_{\alpha_i}^2 \bigg) U_\alpha \bar{U}_\alpha, \quad |U|_{\pi,k} = \sqrt{\langle U, U \rangle_{\pi,k}}. \qquad (2.41)$$

If π_i are constant, then $\big|D^k F\big|_{\pi,k} = |D^{\pi,k} F|$ but, in general, this is false. Moreover, we define

$$\|F\|_{1,\pi,l} = \sum_{k=1}^{l} |D^k F|_{\pi,k}, \quad \|F\|_{\pi,l} = |F| + \|F\|_{1,\pi,l} = \sum_{k=0}^{l} |D^k F|_{\pi,k}. \qquad (2.42)$$

Lemma 2.1.11. *Let*

$$\|\pi\|_{k,\infty} := 1 \vee \max_{i=1,\ldots,J} \max_{|\alpha| \leq k} \|\partial_\alpha \pi_i\|_\infty.$$

For each $k \in \mathbb{N}$ there exists a universal constant C_k such that

$$|D^{\pi,k}F| \leq C_k \|\pi\|_{k-1,\infty}^k |D^k F|_{\pi,k}. \tag{2.43}$$

As an immediate consequence

$$|F|_{\pi,l} \leq C_l \|\pi\|_{l-1,\infty}^l \|F\|_{\pi,l}, \quad |L^\pi F|_{\pi,l} \leq C_l \|\pi\|_{2l-1,\infty}^{2l} \|LF\|_{\pi,l}. \tag{2.44}$$

The proof is straightforward, so we skip it.

We also express the Malliavin covariance matrix in terms of the above scalar product:

$$\sigma_\pi^{i,j}(F) = \langle DF^i, D^j F\rangle_\pi$$

and, similarly to (2.20), we define

$$m_\pi(\sigma_F) = \max\left(1, \frac{1}{\det \sigma_\pi(F)}\right). \tag{2.45}$$

By using the estimate in Lemma 2.1.11, we rewrite Theorem 2.1.10 in terms of the new norms.

Theorem 2.1.12. (i) *For $F \in \mathcal{S}^d$, $G \in \mathcal{S}$, and for all $q \in \mathbb{N}^*$, there exists a universal constant $C_{q,d}$ such that for every multiindex $\beta = (\beta_1, \ldots, \beta_q)$*

$$\left|H_\beta^q(F,G)\right| \leq C_{q,d} B_q(F)^q \|G\|_{\pi,q}, \tag{2.46}$$

with

$$B_q(F) = m_\pi(\sigma_F)^{q+1} \pi_{2q-1,\infty}^{2d(q+1)(q+3)} \left(1 + \|F\|_{\pi,q+1}^{2d(q+2)} + \|LF\|_{\pi,q-1}\right).$$

(ii) *For $F, \overline{F} \in \mathcal{S}^d$, $G, \overline{G} \in \mathcal{S}$, and for all $q \in \mathbb{N}^*$, there exists a universal constant $C_{q,d}$ such that for every multiindex $\beta = (\beta_1, \ldots, \beta_q)$*

$$\left|H_\beta^q(F,G) - H_\beta^q(\overline{F},\overline{G})\right| \leq C_{q,d} B_q(F,\overline{F})^{\frac{q(q+1)}{2}} \left(1 + \|G\|_{\pi,q} + \|\overline{G}\|_{\pi,q}\right)^q$$
$$\times \left(\|F-\overline{F}\|_{\pi,q+1} + \|L(F-\overline{F})\|_{\pi,q-1} + \|G-\overline{G}\|_{\pi,q}\right), \tag{2.47}$$

with

$$B_q(F,\overline{F}) = m_\pi(\sigma_F)^{q+1} m_\pi(\sigma_{\overline{F}})^{q+1} \pi_{2q-1,\infty}^{2d(q+1)(q+4)}$$
$$\times \left(1 + \|F\|_{\pi,q+1}^{2d(q+3)} + \|\overline{F}\|_{\pi,q+1}^{2d(q+3)} + \|LF\|_{\pi,q-1} + \|L\overline{F}\|_{\pi,q-1}\right).$$

2.2   Short introduction to Malliavin calculus

The Malliavin calculus is a stochastic calculus of variations on the Wiener space. Roughly speaking, the strategy to construct it is the following: a class of "regular functionals" on the Wiener space which represent the analogous of the test functions in Schwartz distribution theory is considered. Then the differential operators (which are the analogs of the usual derivatives and of the Laplace operator from the finite-dimensional case) on these regular functionals are defined. These are unbounded operators. Then a duality formula is proven which allows one to check that these operators are closable and, finally, a standard extension of these operators is taken. There are two main approaches: one is based on the expansion in chaos of the functionals on the Wiener space —in this case the "regular functionals" are the multiple stochastic integrals and one defines the differential operators for them. A nice feature of this approach is that it permits to describe the domains of the differential operators (so to characterize the regularity of a functional) in terms of the rate of convergence of the stochastic series in the chaos expansion of the functional. We do not deal with this approach here (the interested reader can consult the classical book Nualart [41]). The second approach (that we follow here) is to take as "regular functionals" the cylindrical functionals —the so-called simple functionals. The specific point in this approach is that the infinite-dimensional Malliavin calculus appears as a natural extension of some finite-dimensional calculus: in a first stage, the calculus in a finite and fixed dimension is established (as presented in the previous section); then, one proves that the definition of the operators do not depend on the dimension and so, a calculus in a finite but arbitrary dimension is obtained. Finally, we pass to the limit and the operators are closed. In the case of the Wiener space the two approaches mentioned above are equivalent (they describe the same objects), but if they are employed on the Poisson space, they lead to different objects obtained. One may consult, e.g., the book by Di Nunno, Øksendal, and Proske [22] on this topic.

2.2.1   Differential operators

On a probability space $(\Omega, \mathcal{F}, P)$ we consider a d-dimensional Brownian motion $W = (W^1, \ldots, W^d)$ and we look at functionals of the trajectories of W. The typical one is a diffusion process. We will restrict ourself to the time interval $[0, 1]$. We proceed in three steps:

Step 1: Finite-dimensional differential calculus in dimension n.

Step 2: Finite-dimensional differential calculus in arbitrary dimension.

Step 3: Infinite-dimensional calculus.

Step 1: Finite-dimensional differential calculus in dimension n

We consider the space of cylindrical functionals and we define the differential operators for them. This is a particular case of the finite-dimensional Malliavin calculus that we presented in Subsection 2.1.1. In a further step we consider the extension of these operators to general functionals on the Wiener space.

We fix n and denote

$$t_n^k = \frac{k}{2^n} \quad \text{and} \quad I_n^k = (t_n^{k-1}, t_n^k], \quad \text{with } k = 1, \ldots, 2^n,$$

and

$$\Delta_n^{k,i} = W^i(t_n^k) - W^i(t_n^{k-1}), \quad i = 1, \ldots, d,$$

$$\Delta_n^k = (\Delta_n^{k,1}, \ldots, \Delta_n^{k,d}), \quad \Delta_n = (\Delta_n^1, \ldots, \Delta_n^{2^n}).$$

We define the simple functionals of order n to be the random variables which depend, in a smooth way, on $W(t_n^k)$ only. More precisely let

$$\mathcal{S}_n = \{F = \phi(\Delta_n) : \phi \in C_p^\infty(\mathbb{R}^{2^n})\},$$

where C_p^∞ denotes the space of infinitely-differentiable functions which have polynomial growth together with their derivatives of any order. We define the simple processes of order n by

$$\mathcal{P}_n = \left\{ U, \ U(s) = \sum_{k=1}^{2^n} 1_{I_n^k}(s) U_k, \quad U_k \in \mathcal{S}_n, \ k = 1, \ldots, 2^n \right\}.$$

We think about the simple functionals as forming a subspace of $L^2(\mathbb{P})$, and about the simple processes as forming a subspace of $L^2(\mathbb{P}; L^2([0,1]))$ (here, $L^2([0,1])$ is the L^2 space with respect to the Lebesgue measure ds and $L^2(P : L^2([0,1]))$ is the space of the square integrable $L^2([0,1])$-valued random variables). This is not important for the moment, but will be crucial when we settle the infinite-dimensional calculus.

Then we define the operators

$$D \colon \mathcal{S}_n \longrightarrow \mathcal{P}_n, \quad \delta \colon \mathcal{P}_n \longrightarrow \mathcal{S}_n$$

as follows. If $F = \phi(\Delta_n)$, then

$$D_s^i F = \sum_{k=1}^{2^n} 1_{I_n^k}(s) \frac{\partial \phi}{\partial \Delta_n^{k,i}}(\Delta_n), \quad i = 1, \ldots, d. \tag{2.48}$$

The link with the Malliavin derivatives defined in Subsection 2.1.1 is the following (we shall be more precise later). The underlying noise is given here by $V = \Delta_n = (\Delta_n^{k,i})_{i=1,\ldots,d, k=1,\ldots,2^n}$ and, with the notations in Subsection 2.1.1,

$$D_s^i F = D_{k,i} F$$

holds for $s \in I_n^k$. Actually, the notation $D^i F$ may be confused with the one we used for iterated derivatives: here, the superscript i relates to the i-th coordinate of the Brownian motion, and not to the order of derivatives. But this is the standard notation in the Malliavin calculus on the Wiener space. So we use it, hoping to be able to be clear enough.

Now we take $U(s) = \sum_{k=1}^{2^n} 1_{I_n^k}(s) U_k$ with $U_k = u_k(\Delta_n)$ and set

$$\delta_i(U) = \sum_{k=1}^{2^n} \left(u_k(\Delta_n)\Delta_n^{k,i} - \frac{\partial u_k}{\partial \Delta_n^{k,i}}(\Delta_n)\frac{1}{2^n} \right). \tag{2.49}$$

For $U = (U^1, \dots, U^d) \in \mathcal{P}_n^d$ we define

$$\delta(U) = \sum_{i=1}^{d} \delta_i(U^i).$$

The typical example of simple process is $U^i(s) = D_s^i F, s \in [0,1]$. It is clear that this process *is not adapted* to the natural filtration associated to the Brownian motion. Suppose however that we have a simple process U which is adapted and let $s \in I_n^k$. Then $u_k(\Delta_n)$ does not depend on Δ_n^k (because $\Delta_n^k = W(t_n^k) - W(t_n^{k-1})$ depends on $W(t_n^k)$ and $t_n^k > s$). It follows that $\partial u_k / \partial\Delta_n^{k,i}(\Delta_n) = 0$. So, for adapted processes, we obtain

$$\delta_i(U) = \sum_{k=1}^{2^n} u_k(\Delta_n)\Delta_n^{k,i} = \int_0^1 U(s)dW_s^i,$$

where the above integral is the Itô integral (in fact it is a Riemann type sum, but, since U is a step function, it coincides with the stochastic integral). This shows that $\delta_i(U)$ is a generalization of the Itô integral for anticipative processes. It is also called the Skorohod integral. This is why in the sequel we will also use the notation

$$\delta_i(U) = \int_0^1 U(s)\widetilde{d}W_s^i.$$

The link between D^i and δ_i is given by the following basic duality formula: for each $F \in \mathcal{S}_n$ and $U \in \mathcal{P}_n$,

$$\mathbb{E}\left(\int_0^1 D_s^i F \times U_s ds \right) = \mathbb{E}\left(F\delta_i(F) \right). \tag{2.50}$$

Notice that this formula may also be written as

$$\left\langle D^i F, U \right\rangle_{L^2(\Omega; L^2(0,1))} = \left\langle F, \delta_i(U) \right\rangle_{L^2(\Omega)}. \tag{2.51}$$

This formula can be easily obtained using integration by parts with respect to the Lebesgue measure. In fact, we have already proved it in Subsection 2.1.2: we have to take the random vector Δ_n instead of the random vector V there (we come back later on the translation of the calculus given for V there and for Δ_n here). The key point is that $\Delta_n^{k,i}$ is a centered Gaussian random variable of variance $h_n = 2^{-n}$, so it has density

$$p_{\Delta_n^{k,i}}(y) = \frac{1}{\sqrt{2\pi h_n}} e^{-y^2/2h_n}.$$

Then

$$\left(\ln p_{\Delta_n^{k,i}}\right)'(y) = -\frac{y}{h_n}$$

so, (2.50) is a particular case of (2.4).

We define now higher-order derivatives. Let $\alpha = (\alpha_1, \ldots, \alpha_r) \in \{1, \ldots, d\}^N$ and recall that $|\alpha| = r$ denotes the length of the multiindex α. Define

$$D^{\alpha}_{s_1, \ldots, s_r} F = D^{\alpha_r}_{s_r} \cdots D^{\alpha_1}_{s_1} F.$$

This means that if $s_p \in I_n^{k_p}, p = 1, \ldots, r$, then

$$D^{\alpha}_{s_1, \ldots, s_r} F = \frac{\partial^r}{\partial \Delta_n^{k_r, \alpha_r} \cdots \partial \Delta_n^{k_1, \alpha_1}} \phi(\Delta_n).$$

We regard $D^{\alpha} F$ as an element of $L^2(\mathbb{P}; L^2([0, 1]^r))$.

Let us give the analog of (2.50) for iterated derivatives. For $U_i \in \mathcal{P}_n$, $i = 1, \ldots, r$, we can define by recurrence

$$\int_0^1 \tilde{d}W^{\alpha_1}_{s_1} \cdots \int_0^1 \tilde{d}W^{\alpha_r}_{s_r} U_1(s_1) \cdots U_r(s_r).$$

Indeed, for each fixed $s_1, \ldots, s_{r-1}$ the process $s_r \mapsto U_1(s_1) \cdots U_r(s_r)$ is an element of $\mathcal{P}_n$ and we can define $\delta_{\alpha_r}(U_1(s_1) \cdots U_{r-1}(s_{r-1})U_r(\circ))$. And we continue in the same way. Once we have defined this object, we use (2.50) in a recurrent way and we obtain

$$\mathbb{E}\left(\int_{(0,1)^r} D^{\alpha i}_{s_1, \ldots, s_r} F \times U_1(s_1) \cdots U_r(s_r) ds_1 \cdots ds_r\right)$$

$$= \mathbb{E}\left(F \int_0^1 \tilde{d}W^{\alpha_1}_{s_1} \cdots \int_0^1 \tilde{d}W^{\alpha_r}_{s_r} U_1(s_1) \cdots U_r(s_r)\right).$$

(2.52)

Finally, we define $L \colon \mathcal{S} \to \mathcal{S}$ by

$$L(F) = \sum_{i=1}^d \delta_i(D^i F).$$

(2.53)

This is the so-called Ornstein–Uhlenbeck operator (we do not explain here the reason for their terminology; see Nualart [41] or Sanz-Solé [44]). The duality property for L is

$$\mathbb{E}(FLG) = \mathbb{E}(\langle DF, DG \rangle) = \mathbb{E}(GLF),$$

(2.54)

in which we have set

$$\mathbb{E}(\langle DF, DG\rangle) = \sum_{i=1}^{d} \mathbb{E}(\langle D^i F, D^i G\rangle).$$

Step 2: Finite-dimensional differential calculus in arbitrary dimension

We define the general simple functionals and the general simple processes by

$$\mathcal{S} = \bigcup_{n=1}^{\infty} \mathcal{S}_n \quad \text{and} \quad \mathcal{P} = \bigcup_{n=1}^{\infty} \mathcal{P}_n.$$

Then we define $D\colon \mathcal{S} \to \mathcal{P}$ and $\delta\colon \mathcal{P} \to \mathcal{S}$ in the following way: if $F \in \mathcal{S}_n$, then DF is given by the formula (2.48), and if $U \in \mathcal{P}_n$, then $\delta_i(U)$ is given by the formula (2.49). Notice that $\mathcal{S}_n \subset \mathcal{S}_{n+1}$. So we have to make sure that, using the definition of differential operator given in (2.48) regarding F as an element of $\mathcal{S}_n$ or of $\mathcal{S}_{n+1}$, we obtain the same thing. And it is easy to check that this is the case, so the definition given above is correct. The same is true for δ_i.

Now we have D and δ and it is clear that the duality formulas (2.50) and (2.52) hold for $F \in \mathcal{S}$ and $U \in \mathcal{P}$.

Step 3: Infinite-dimensional calculus

At this point, we have the unbounded operators

$$D\colon \mathcal{S} \subset L^2(\Omega) \longrightarrow \mathcal{P} \subset L^2(\Omega; L^2(0,1))$$

and

$$\delta_i\colon \mathcal{P} \subset L^2(\Omega; L^2(0,1)) \longrightarrow \mathcal{S} \subset L^2(\Omega)$$

defined above. We also know (we do not prove it here) that $\mathcal{S} \subset L^2(\Omega)$ and $\mathcal{P} \subset L^2(\Omega; L^2(0,1))$ are dense. So we follow the standard procedure for extending unbounded operators. The first step is to check that they are closable:

Lemma 2.2.1. *Let $F \in L^2(\Omega)$ and $F_n \in \mathcal{S}$, $n \in \mathbb{N}$, be such that $\lim_n \|F_n\|_{L^2(\Omega)} = 0$ and*

$$\lim_n \|D^i F_n - U_i\|_{L^2(\Omega; L^2(0,1))} = 0$$

for some $U_i \in L^2(\Omega; L^2(0,1))$. Then, $U_i = 0$.

Proof. We fix $\bar{U} \in \mathcal{P}$ and use the duality formula (2.51) in order to obtain

$$\mathbb{E}\big(\langle U_i, \bar{U}\rangle_{L^2(0,1)}\big) = \lim_n \mathbb{E}\big(\langle D^i F_n, \bar{U}\rangle_{L^2(0,1)}\big) = \lim_n \mathbb{E}\big(F_n \delta_i(\bar{U})\big) = 0.$$

Since $\mathcal{P} \subset L^2(\Omega; L^2(0,1))$ is dense, we conclude that $U_i = 0$. $\qquad\square$

We are now able to introduce

Definition 2.2.2. *Let $F \in L^2(\Omega)$. Suppose that there exists a sequence of simple functionals $F_n \in \mathcal{S}$, $n \in \mathbb{N}$, such that $\lim_n F_n = F$ in $L^2(\Omega)$ and $\lim_n D^i F_n = U_i$ in $L^2(\Omega; L^2(0,1))$. Then we say that $F \in \operatorname{Dom} D^i$ and we define $D^i F = U_i = \lim_n D^i F_n$.*

Notice that the definition is correct: it does not depend on the sequence F_n, $n \in \mathbb{N}$. This is because D^i is closable. This definition is the analog of the definition of the weak derivatives in L^2 sense in the finite-dimensional case. So, $\operatorname{Dom} D := \cap_{i=1}^d \operatorname{Dom} D^i$ is the analog of $H^1(\mathbb{R}^m)$ if we work on the finite-dimensional space $\mathbb{R}^m$. In order to be able to use Hölder inequalities we have to define derivatives in L^p sense for any $p > 1$. This will be the analog of $W^{1,p}(\mathbb{R}^m)$.

Definition 2.2.3. *Let $p > 1$ and let $F \in L^p(\Omega)$. Suppose that there exists a sequence of simple functionals F_n, $n \in \mathbb{N}$, such that $\lim_n F_n = F$ in $L^p(\Omega)$ and $\lim_n D^i F_n = U_i$ in $L^p(\Omega; L^2(0,1))$, $i = 1, \ldots, d$. Then we say that $F \in D^{1,p}$ and we define $D^i F = U_i = \lim_n D^i F_n$.*

Notice that we still look at $D^i F_n$ as at an element of $L^2(0,1)$. The power p concerns the expectation only. Notice also that $\mathcal{P} \subset L^p(\Omega; L^2(0,1))$ is dense for $p > 1$, so the same reasoning as above allows one to prove that D^i is closable as an operator defined on $\mathcal{S} \subset L^p(\Omega)$.

We introduce now the Sobolev-like norm:

$$|F|_{1,1} = \sum_{i=1}^d \left\| D^i F \right\|_{L^2(0,1)} = \sum_{i=1}^d \left(\int_0^1 |D_s^i F|^2 \, ds \right)^{1/2} \quad \text{and} \quad |F|_1 = |F| + |F|_{1,1}.$$

(2.55)

Moreover, we define

$$\|F\|_{1,p} = \|F\|_p + \sum_{i=1}^d \left\| \left\| D^i F \right\|_{L^2(0,1)} \right\|_p$$

(2.56)

$$= \left(\mathbb{E}(|F|^p) \right)^{1/p} + \sum_{i=1}^d \left(\mathbb{E}\left(\left(\int_0^1 |D_s^i F|^2 \, ds \right)^{p/2} \right) \right)^{1/p}.$$

It is easy to check that $\mathbb{D}^{1,p}$ is the closure of $\mathcal{S}$ with respect to $\|\circ\|_{1,p}$.

We go now further to the derivatives of higher order. Using (2.52) —instead of (2.50)— it can be checked in the same way as before that, for every multiindex $\alpha = (\alpha_1, \ldots, \alpha_r) \in \{1, \ldots, d\}^r$, the operator D^α is closable. So we may give the following

Definition 2.2.4. *Let $p > 1$ and let $F \in L^p(\Omega)$. Consider a multiindex $\alpha = (\alpha_1, \ldots, \alpha_r) \in \mathbb{N}^d$. Suppose that there exists a sequence of simple functionals F_n, $n \in \mathbb{N}$, such that $\lim_n F_n = F$ in $L^p(\Omega)$ and $\lim_n D^\alpha F_n = U_i$ in $L^p(\Omega; L^2((0,1)^r))$. Then we define $D^i F = U_i = \lim_n D^i F_n$. We denote by $\mathbb{D}^{k,p}$ the class of functionals F for which D^α defined above exists for every multiindex α with $|\alpha| \leq k$.*

We introduce the norms

$$|F|_{1,k} = \sum_{r=1}^{k} \sum_{|\alpha|=r} \|D^{\alpha}F\|_{L^2((0,1)^r)}, \quad |F|_k = |F| + |F|_{1,k} \tag{2.57}$$

and

$$\|F\|_{k,p} = \left(\mathbb{E}(|F|_k^p)\right)^{1/p} = \|F\|_p + \sum_{r=1}^{k} \sum_{|\alpha|=r} \left\| \|D^{\alpha}F\|_{L^2((0,1)^r)} \right\|_p \tag{2.58}$$

$$= \left(\mathbb{E}(|F|^p)\right)^{1/p} + \sum_{r=1}^{k} \sum_{|\alpha|=r} \left(\mathbb{E}\left(\left(\int_{(0,1)^r} \left|D^{\alpha}_{s_1,\dots,s_r}F\right|^2 ds_1 \cdots ds_r\right)^{p/2}\right)\right)^{1/p}.$$

It is easy to check that $\mathbb{D}^{k,p}$ is the closure of $\mathcal{S}$ with respect to $\|\circ\|_{k,p}$. Clearly,

$$\mathbb{D}^{k,p} \subset \mathbb{D}^{k',p'} \quad \text{if} \quad k \geq k', \ p \geq p'.$$

We define

$$\mathbb{D}^{k,\infty} = \bigcap_{p>1} \mathbb{D}^{k,p}, \quad \mathbb{D}^{\infty} = \bigcap_{k=1}^{\infty} \bigcap_{p>1} \mathbb{D}^{k,p}.$$

Notice that the correct notation would be $\mathbb{D}^{k,\infty-}$ and $\mathbb{D}^{\infty-}$ in order to avoid the confusion with the possible use of $\|\circ\|_{\infty}$; for simplicity, we shall omit "$-$".

We turn now to the extension of δ_i. This operator is closable (the same argument as above can be applied, based on the duality relation (2.50) and on the fact that $\mathcal{S} \subset L^p(\Omega)$ is dense). So we may give the following

Definition 2.2.5. *Let $p > 1$ and let $U \in L^p(\Omega; L^2(0,1))$. Suppose that there exists a sequence of simple processes $U_n \in \mathcal{P}$, $n \in \mathbb{N}$, such that $\lim_n U_n = UF$ in $L^p(\Omega; L^2(0,1))$ and $\lim_n \delta_i(U_n) = F$ in $L^p(\Omega)$. In this situation, we define $\delta_i(U) = \lim_n \delta_i(U_n)$. We denote by $\mathrm{Dom}_p(\delta)$ the space of those processes $U \in L^p(\Omega; L^2(0,1))$ for which the above property holds for every $i = 1, \dots, d$.*

In a similar way we may define the extension of L.

Definition 2.2.6. *Let $p > 1$ and let $F \in L^p(\Omega)$. Suppose that there exists a sequence of simple functionals $F_n \in \mathcal{S}$, $n \in \mathbb{N}$, such that $\lim_n F_n = F$ in $L^p(\Omega)$ and $\lim_n LF_n = F$ in $L^p(\Omega)$. Then we define $LF = \lim_n LF_n$. We denote by $\mathrm{Dom}_p(L)$ the space of those functionals $F \in L^p(\Omega)$ for which the above property holds.*

These are the basic operators in Malliavin calculus. There exists a fundamental result about the equivalence of norms due to P. Meyer (and known as Meyer's inequalities) which relates the Sobolev norms defined in (2.58) with another class of norms based on the operator L. We will not prove this nontrivial result here referring the reader to the books Nualart [41] and Sanz-Solé [44].

Theorem 2.2.7. *Let $F \in \mathbb{D}^\infty$. For each $k \in \mathbb{N}$ and $p > 1$ there exist constants $c_{k,p}$ and $C_{k,p}$ such that*

$$c_{k,p} \|F\|_{k,p} \leq \sum_{i=0}^{k} \left\| L^{i/2} F \right\|_p \leq C_{k,p} \|F\|_{k,p} . \tag{2.59}$$

In particular, $\mathbb{D}^\infty \subset \mathrm{Dom}\, L$.

As an immediate consequence of Theorem 2.2.7, for each $k \in \mathbb{N}$ and $p > 1$ there exists a constant $C_{k,p}$ such that

$$\|LF\|_{k,p} \leq C_{k,p} \|F\|_{k+2,p}. \tag{2.60}$$

2.2.2 Computation rules and integration by parts formulas

In this subsection we give the basic computation formulas, the integration by parts formulas (IBP formulas for short) and estimates of the weights which appear in the IBP formulas. We proceed in the following way. We first fix n and consider functionals in $\mathcal{S}_n$. In this case we fit in the framework from Section 2.1. We identify the objects and translate the results obtained therein to the specific context of the Malliavin calculus. And, in a second step, we pass to the limit and obtain similar results for general functionals.

We recall that in Section 2.1 we looked at functionals of the form $F = f(V)$ with $V = (V_1, \ldots, V_J)$, where $J \in \mathbb{N}$. And the law of V is absolutely continuous with respect to the Lebesgue measure and has the density p_J. In our case we fix n and consider $F \in \mathcal{S}_n$. Then we take V to be $\Delta_n^{k,i} = W^i(k/2^n) - W^i((k-1)/2^n)$, $k = 1, \ldots, 2^n$, $i = 1, \ldots, d$. So, $J = d \times 2^n$ and p_J is a Gaussian density. We work with the weights $\pi_i = 2^{-n/2}$. The derivative defined in (2.1) is (with i replaced by (k,i))

$$D_{k,i} F = \frac{1}{2^{n/2}} \frac{\partial F}{\partial \Delta_n^{k,i}}.$$

It follows that the relation between the derivative defined in (2.1) and the one defined in (2.48) is given by

$$D_s^i F = 2^{n/2} D_{k,i} F = \frac{\partial F}{\partial \Delta_n^{k,i}}, \quad \text{for} \quad \frac{k-1}{2^n} \leq s < \frac{k}{2^n}.$$

Coming to the norm defined in (2.14) we have

$$|DF|^2 := \sum_{i=1}^{d} \sum_{k=1}^{2^n} |D_{k,i} F|^2 = \sum_{i=1}^{d} \int_0^1 |D_s^i F|^2 \, ds = \sum_{i=1}^{d} \left\| D^i F \right\|^2_{L^2(0,1)} .$$

We identify in the same way the higher-order derivatives and thus obtain the following expression for the norms defined in (2.15):

$$|F|_{1,l}^2 = \sum_{k=1}^{l} \sum_{|\beta|=k} \left\| D^\beta F \right\|^2_{L^2(0,1)^k}, \quad |F|_l = |F| + |F|_{1,l}.$$

We turn now to the Skorohod integral. According to (2.2), if we take $\pi_{k,i} = 2^{-n/2}$ and $V = \Delta_n$, we obtain for $U = (2^{-n/2}u_{k,i}(\Delta_n))_{k,i}$:

$$\delta(U) = \sum_{i=1}^{d}\sum_{k=1}^{2^n} \delta_{k,i}(U) = -2^{-n}\sum_{i=1}^{d}\sum_{k=1}^{2^n} \left(\partial_{\Delta_n^{k,i}} u_{k,i} + u_{k,i}\partial_{\Delta_n^{k,i}} \ln p_{\Delta_n^{k,i}}\right)(\Delta_n)$$

$$= -2^{-n}\sum_{i=1}^{d}\sum_{k=1}^{2^n} \left(\partial_{\Delta_n^{k,i}} u_{k,i}(\Delta_n) - u_{k,i}(\Delta_n)\frac{\Delta_n^{k,i}}{2^{-n}}\right)$$

$$= \sum_{i=1}^{d}\sum_{k=1}^{2^n} u_{k,i}(\Delta_n)\Delta_n^{k,i} - \sum_{i=1}^{d}\sum_{k=1}^{2^n} \partial_{\Delta_n^{k,i}} u_{k,i}(\Delta_n)\frac{1}{2^n}.$$

It follows that

$$L(DF) = \delta(DF) = \sum_{i=1}^{d}\sum_{k=1}^{2^n} 1_{I_k^i}(s)D_s^i F \times \Delta_n^{k,i} - \sum_{i=1}^{d}\sum_{k=1}^{2^n} 1_{I_k^i}(s)D_s^i D_s^i F \frac{1}{2^n},$$

and this shows that the definition of L given in (2.53) coincides with the one given in (2.3).

We come now to the computation rules. We will extend the properties defined in the finite-dimensional case to the infinite-dimensional one.

Lemma 2.2.8. *Let $\phi\colon \mathbb{R}^d \to \mathbb{R}$ be a smooth function and $F = (F^1, \ldots, F^d) \in (\mathbb{D}^{1,p})^d$ for some $p > 1$. Then $\phi(F) \in \mathbb{D}^{1,p}$ and*

$$D\phi(F) = \sum_{r=1}^{d} \partial_r \phi(F)DF^r. \tag{2.61}$$

If $F \in \mathbb{D}^{1,p}$ and $U \in \mathrm{Dom}\,\delta$, then

$$\delta(FU) = F\delta(U) - \langle DF, U\rangle_{L^2(0,1)}. \tag{2.62}$$

Moreover, for $F = (F^1, \ldots, F^d) \in (\mathrm{Dom}\,L)^d$, we have

$$L\phi(F) = \sum_{r=1}^{d} \partial_r \phi(F)LF^r - \sum_{r,r'=1}^{d} \partial_{r,r'}\phi(F)\langle DF^r, DF^{r'}\rangle_{L^2(0,1)}. \tag{2.63}$$

Proof. If $F \in \mathcal{S}_n^d$, equality (2.61) coincides with (2.5). For general $F \in (\mathbb{D}^{1p})^d$ we take a sequence $F \in \mathcal{S}^d$ such that $F_n \to F$ in $L^p(\Omega)$ and $DF_n \to DF$ in $L^p(\Omega; L^2(0,1))$. Passing to a subsequence, we obtain almost sure convergence and so we may pass to the limit in (2.61). The other equalities are obtained in a similar way. $\quad\square$

For $F \in (\mathbb{D}^{1,p})^d$ we define the Malliavin covariance matrix σ_F by

$$\sigma_F^{i,j} = \langle DF^i, DF^j \rangle_{L^2((0,1);\mathbb{R}^d)} = \sum_{r=1}^{d} \int_0^1 D_s^r F^i D_s^r F^j \, ds, \quad i,j = 1,\ldots,d. \quad (2.64)$$

If $\omega \in \{\det \sigma_F \neq 0\}$, we define $\gamma_F(\omega) = \sigma_F^{-1}(\omega)$. We can now state the main results of this section, the integration by parts formula given by Malliavin.

Theorem 2.2.9. *Let $F = (F^1, \ldots, F^d) \in (\mathbb{D}^{1,\infty})^d$ and suppose that*

$$\mathbb{E}\big(|\det \sigma_F|^{-p}\big) < \infty \quad \forall p \geq 1. \quad (2.65)$$

Let $\phi \colon \mathbb{R}^d \to \mathbb{R}$ be a smooth bounded function with bounded derivatives.

(i) *If $F = (F^1, \ldots, F^d) \in (\mathbb{D}^{2,\infty})^d$ and $G \in \mathbb{D}^{1,\infty}$, then for every $r = 1,\ldots,d$ we have*

$$\mathbb{E}\left(\partial_r \phi(F) G\right) = \mathbb{E}\left(\phi(F) H_r(F,G)\right), \quad (2.66)$$

where

$$H_r(F,G) = \sum_{r'=1}^{d} \delta\big(G\gamma^{r',r}(F)DF^{r'}\big)$$

$$= \sum_{r'=1}^{d} \big(G\delta(\gamma_F^{r',r} DF^{r'}) - \gamma_F^{r',r} \langle DF^{r'}, DG \rangle_{L^2(0,1)}\big). \quad (2.67)$$

(ii) *If $F = (F^1, \ldots, F^d) \in (\mathbb{D}^{q+1,\infty})^d$ and $G \in \mathbb{D}^{q,\infty}$, then, for every multiindex $\alpha = (\alpha_1, \ldots, \alpha_d) \in \{1, \ldots, d\}^q$ we have*

$$\mathbb{E}\big(\partial_\alpha \phi(F) G\big) = \mathbb{E}\big(\phi(F) H_\alpha^q(F,G)\big), \quad (2.68)$$

with $H_\alpha(F,G)$ defined by recurrence:

$$H_{(\alpha,i)}^{q+1}(F,G) = H_i(F, H_\alpha^q(F,G)). \quad (2.69)$$

Proof. We take a sequence F_n, $n \in \mathbb{N}$, of simple functionals such that $F_n \to F$ in $\|\circ\|_{2,p}$ for every $p \in \mathbb{N}$, and a sequence G_n, $n \in \mathbb{N}$, of simple functionals such that $G_n \to G$ in $\|\circ\|_{1,p}$ for every $p \in \mathbb{N}$. We use Theorem 2.1.4 in order to obtain the integration by parts formula (2.66) for F_n and G_n, and then we pass to the limit and we obtain it for F and G. This reasoning has a weak point: we know that the non-degeneracy property (2.65) holds for F but not for F_n (and this is necessary in order to be able to use Theorem 2.1.4). Although this is an unpleasant technical problem, there is no real difficulty here, and we may remove it in the following way: we consider a d-dimensional standard Gaussian random vector U which is independent of the basic Brownian motion (we can construct it on an extension of the probability space) and we define $\overline{F}_n = F_n + \frac{1}{n}U$. We will use the integration by parts formula for the couple Δ_n (the increments of the Brownian motion) and U. Then, the Malliavin covariance matrix will be $\sigma_{\overline{F}_n} = \sigma_{F_n} + \frac{1}{n}I$, where I is the identity matrix. Now $\det \sigma_{\overline{F}_n} \geq n^{-d} > 0$ so we obtain (2.66) for $\overline{F}_n$ and G_n. Then we pass to the limit. So (i) is proved. And (ii) follows by recurrence.   □

Remark 2.2.10. Another way to prove the above theorem is to reproduce exactly the reasoning from the proof of Theorem 2.1.4. This is possible because the only thing which is used there is the chain rule and we have it for general functionals on the Wiener space, see (2.61). In fact, this is the natural way to do the proof, and this is the reason to define an infinite-dimensional differential calculus. We have proposed the proof based on approximation just to illustrate the link between the finite-dimensional calculus and the infinite-dimensional one.

We give now the basic estimates for $H_\beta^q(F, G)$. One can follow the same arguments that allowed to get Theorem 2.1.10. So, for $F \in (\mathbb{D}^{1,2})^d$ we set

$$m(\sigma_F) = \max\left(1, \frac{1}{\det \sigma_F}\right). \tag{2.70}$$

Theorem 2.2.11.   (i) *Let* $q \in \mathbb{N}^*$, $F \in (\mathbb{D}^{q+1,\infty})^d$, *and* $G \in \mathbb{D}^{q,\infty}$. *There exists a universal constant* $C_{q,d}$ *such that, for every multiindex* $\beta = (\beta_1, \ldots, \beta_q)$,

$$\left|H_\beta^q(F, G)\right| \leq C_{q,d} A_q(F)^q |G|_q, \tag{2.71}$$

with

$$A_q(F) = m(\sigma_F)^{q+1} \left(1 + |F|_{q+1}^{2d(q+2)} + |LF|_{q-1}\right).$$

Here, we make the convention $A_q(F) = \infty$ *if* $\det \sigma_F = 0$.

(ii) *Let* $q \in \mathbb{N}^*$, $F, \overline{F} \in (\mathbb{D}^{q+1,\infty})^d$, *and* $G, \overline{G} \in \mathbb{D}^{q,\infty}$. *There exists a universal constant* $C_{q,d}$ *such that, for every multiindex* $\beta = (\beta_1, \ldots, \beta_q)$,

$$\left|H_\beta^q(F, G) - H_\beta^q(\overline{F}, \overline{G})\right| \leq C_{q,d} A_q(F, \overline{F})^{\frac{q(q+1)}{2}} \times \left(1 + |G|_q + |\overline{G}|_q\right)^q$$
$$\times \left(\left|F - \overline{F}\right|_{q+1} + \left|L(F - \overline{F})\right|_{q-1} + \left|G - \overline{G}\right|_q\right), \tag{2.72}$$

with

$$A_q(F, \overline{F}) = m(\sigma_F)^{q+1} m(\sigma_{\overline{F}})^{q+1} \left(1 + |F|_{q+1}^{2d(q+3)} + |\overline{F}|_{q+1}^{2d(q+3)} + |LF|_{q-1} + |L\overline{F}|_{q-1}\right).$$

Proof. This is obtained by passing to the limit in Theorem 2.1.10: we take a sequence F_n, $n \in \mathbb{N}$, of simple functionals such that $F_n \to F$, $D^\alpha F_n \to D^\alpha F$, and $LF_n \to LF$ almost surely (knowing that all the approximations hold in L^p, we may pass to a subsequence and obtain almost sure convergence). We also take a sequence G_n, $n \in \mathbb{N}$, of simple functionals such that $G_n \to G$ and $D^\alpha G_n \to D^\alpha G$ almost surely. Then, we use Theorem 2.1.10 to obtain

$$|H^q_\beta(F_n, G_n)| \le C_{q,d} A_q(F_n)^q |G_n|_q.$$

Notice that we may have $\det \sigma_{F_n} = 0$, but in this case we put $A_q(F_n) = \infty$ and the inequality still holds. Then we pass to the limit and we obtain $|H^q_\beta(F, G)| \le C_{q,d} A_q(F)^q |G|_q$. Here it is crucial that the constants $C_{q,d}$ given in Theorem 2.1.10 are universal constants (which do not depend on the dimension J of the random vector V involved in that theorem). □

So, the estimate from Theorem 2.2.11 gives bounds for the L^p norms of the weights and for the difference between two weights, as it immediately follows by applying the Hölder and Meyer inequalities (2.60). We resume the results in the following corollary.

Corollary 2.2.12. (i) *Let $q, p \in \mathbb{N}^*$, $F \in (\mathbb{D}^{q+1,\infty})^d$, and $G \in \mathbb{D}^{q,\infty}$ such that $(\det \sigma_F)^{-1} \in \cap_{p>1} L^p$. There exist two universal constants C, l (depending on p, q, d only) such that, for every multiindex $\beta = (\beta_1, \ldots, \beta_q)$,*

$$\|H^q_\beta(F, C)\|_p \le C \, B_{q,l}(F)^q \, \|G\|_{q,l}, \tag{2.73}$$

where

$$B_{q,l}(F) = \left(1 + \|(\det \sigma_F)^{-1}\|_l\right)^{q+1} \left(1 + \|F\|_{q+1,l}\right)^{2d(q+2)}. \tag{2.74}$$

(ii) *Let $q, p \in \mathbb{N}^*$, $F, \overline{F} \in (\mathbb{D}^{q+1,\infty})^d$, and $G, \overline{G} \in \mathbb{D}^{q,\infty}$ be such that $(\det \sigma_F)^{-1} \in \cap_{p>1} L^p$ and $(\det \sigma_{\overline{F}})^{-1} \in \cap_{p>1} L^p$. There exists two universal constants C, l (depending on p, q, d only) such that, for every multiindex $\beta = (\beta_1, \ldots, \beta_q)$,*

$$\|H^q_\beta(F, G) - H^q_\beta(\overline{F}, \overline{G})\|_p \le C \, B_{q,l}(F, \overline{F})^{\frac{q(q+1)}{2}} \left(1 + \|G\|^q_{q,l} + \|\overline{G}\|^q_{q,l}\right)$$
$$\times \left(\|F - \overline{F}\|_{q+1,l} + \|G - \overline{G}\|_{q+1,l}\right), \tag{2.75}$$

where

$$B_{q,l}(F, \overline{F}) = \left(1 + \|(\det \sigma_F)^{-1}\|_l + \|(\det \sigma_{\overline{F}})^{-1}\|_l\right)^{2(q+1)}$$
$$\left(1 + \|F\|_{q+1,l} + \|\overline{F}\|_{q+1,l}\right)^{2d(q+3)}. \tag{2.76}$$

We introduce now a localization random variable. Let $m \in \mathbb{N}$ and let $\phi \in C_p^\infty(\mathbb{R}^m)$. We denote by A_ϕ the complement of the set $\text{Int}\{\phi = 0\}$. Notice that the derivatives of ϕ are null on $A_\phi^c = \text{Int}\{\phi = 0\}$. Consider now $\Theta \in (\mathbb{D}^\infty)^m$ and let $V = \phi(\Theta)$. Clearly, $V \in \mathbb{D}^\infty$. The important point is that

$$V = 1_{A_\phi}(\Theta)V \quad \text{and} \quad D^\alpha V = 1_{A_\phi}(\Theta)D^\alpha V \tag{2.77}$$

for every multiindex α. Coming back to (2.67) and (2.69), we get

$$H_\beta^q(F, G\phi(\Theta)) = 1_{A_\phi}(\Theta) H_\beta^q(F, G\phi(\Theta)). \tag{2.78}$$

So, in (2.71) we may replace $A_q(F)$ and G with $1_{A_\phi}(\Theta) A_q(F)$ and $G\phi(\Theta)$, respectively. Similarly, in (2.72) we may replace $A_q(F, \overline{F})$ and G with $1_{A_\phi}(\Theta) A_q(F, \overline{F})$ and $G\phi(\Theta)$, respectively. And, by using again the Hölder and Meyer inequalities, we immediately get

Corollary 2.2.13.   (i) *Let $q, p \in \mathbb{N}^*$, $F \in (\mathbb{D}^{q+1,\infty})^d$, $G \in \mathbb{D}^{q,\infty}$, $\phi \in C_p^\infty(\mathbb{R}^m)$ and $\Theta \in (\mathbb{D}^\infty)^m$. We denote $V = \phi(\Theta)$. We assume that $1_{A_\phi}(\Theta)(\det \sigma_F)^{-1} \in \cap_{p>1} L^p$. There exists some universal constant C, l (depending on p, q, d only), such that, for every multiindex $\beta = (\beta_1, \ldots, \beta_q)$,*

$$\left\| H_\beta^q(F, GV) \right\|_p \le C B_{q,l}^V(F)^q \|G\|_{q,l}, \tag{2.79}$$

where

$$B_{q,l}^V(F) = \left(1 + \|1_{A_\phi}(\Theta)(\det \sigma_F)^{-1}\|_l\right)^{q+1} \|\phi(\Theta)\|_{q,l} \left(1 + \|F\|_{q+1,l}\right)^{2d(q+2)}. \tag{2.80}$$

(ii) *Let $q, p \in \mathbb{N}^*$, $F, \overline{F} \in (D^{q+1,\infty})^d$, and $G, \overline{G} \in D^{q,\infty}$ and assume that $1_{A_\phi}(\Theta)(\det \sigma_F)^{-1} \in \cap_{r>1} L^r$ and $1_{A_\phi}(\Theta)(\det \sigma_{\overline{F}})^{-1} \in \cap_{r>1} L^r$. There exists some universal constants C, l (depending on p, q, d only) such that, for every multiindex $\beta = (\beta_1, \ldots, \beta_q)$,*

$$\|H_\beta^q(F, GV) - H_\beta^q(\overline{F}, \overline{G}V)\|_p \le C \, B_{q,l}^V(F, \overline{F})^{\frac{q(q+1)}{2}} \left(1 + \|G\|_{q,l}^q + \|\overline{G}\|_{q,l}^q\right)$$
$$\times \left(\|F - \overline{F}\|_{q+1,l} + \|G - \overline{G}\|_{q+1,l}\right), \tag{2.81}$$

where

$$B_{q,l}^V(F, \overline{F}) = \left(1 + \|1_{A_\phi}(\Theta)(\det \sigma_F)^{-1}\|_l + \|1_{A_\phi}(\Theta)(\det \sigma_{\overline{F}})^{-1}\|_l\right)^{2(q+1)}$$
$$\times \left(1 + \|\Phi(\Theta)\|_{q+1,l}\right)^2 \left(1 + \|F\|_{q+1,l} + \|\overline{F}\|_{q+1,l}\right)^{2d(q+3)}. \tag{2.82}$$

Notice that we used the localization $1_{A_\phi}(\Theta)$ for $(\det \sigma_F)^{-1}$, but not for $\|F\|_{q+1,p}$. This is because we used Meyer's inequality to bound $\|1_{A_\phi}(\Theta) LF\|_{q-1,p}$ from above, and this is not possible if we keep $1_{A_\phi}(\Theta)$.

2.3   Representation and estimates for the density

In this section we give the representation formula for the density of the law of a Wiener functional using the Poisson kernel Q_d and, moreover, we give estimates of the tails and of the densities. These are immediate applications of the general results presented in Section 1.8. We first establish the link between the estimates from the previous sections and the notation in 1.8. We recall that there we used the norm $\|G\|_{W_F^{q,p}}$, and in Remark 1.8.2 we gave an upper bound for this quantity in terms of the weights $\|H_\alpha^q(F, G)\|_p$. Combining this with (2.73), we have the following: for each $q \in \mathbb{N}$, $p \geq 1$, there exists some universal constants C, l, depending only on d, q, p, such that

$$\|G\|_{W_F^{q,p}} \leq C B_{q,l}(F)^q \|G\|_{q,l}, \tag{2.83}$$

where $B_{q,l}(F)$ is defined by (2.74). And in the special case $G = 1$ and $q = 1$, we get

$$\|1\|_{W_F^{1,p}} \leq C B_{1,l}(F), \tag{2.84}$$

for suitable $C, l > 0$ depending on p and d. This allows us to use Theorems 1.8.3 and 1.5.2, to derive the following result.

Theorem 2.3.1. (i) *Let* $F = (F^1, \ldots, F^d) \in (\mathbb{D}^{2,\infty})^d$ *be such that, for every* $p > 0$, $\mathbb{E}(|\det \sigma_F|^{-p}) < \infty$. *Then, the law of* F *is absolutely continuous with respect to the Lebesgue measure and has a continuous density* $p_F \in C(\mathbb{R}^d)$ *which is represented as*

$$p_F(x) = \sum_{i=1}^{d} \mathbb{E}(\partial_i Q_d(F - x) H_i(F, 1)), \tag{2.85}$$

where $H_i(F, 1)$ *is defined as in (2.67). Moreover, with* $p > d$ *and* $k_{p,d} = (d-1)/(1 - d/p)$, *there exist* $C > 0$ *and* $p' > d$ *depending on* p, d, *such that*

$$\mathbb{E}(|\nabla Q_d(F - x)|^{\frac{p}{p-1}})^{\frac{p-1}{p}} \leq C B_{1,p'}(F)^{k_{p,d}}, \tag{2.86}$$

where $B_{1,p'}(F)$ *is given by (2.74).*

(ii) *With* $p > d$, $k_{p,d} = (d-1)/(1 - d/p)$, *and* $0 < a < 1/d - 1/p$, *there exist* $C > 0$ *and* $p' > d$ *depending on* p, d, *such that*

$$p_F(x) \leq C B_{1,p'}(F)^{1+k_{p,d}} \mathbb{P}(F \in B_2(x))^a. \tag{2.87}$$

(iii) *If* $F \in (\mathbb{D}^{q+2,\infty})^d$, *then* $p_F \in C^q(\mathbb{R}^d)$ *and, for* $|\alpha| \leq q$, *we have*

$$\partial_\alpha p_F(x) = \sum_{i=1}^{d} \mathbb{E}(\partial_i Q_d(F - x) H_{(i,\alpha)}(F, 1)).$$

Moreover, with $p > d$, $k_{p,d} = (d-1)/(1 - d/p)$, *and* $0 < a < 1/d - 1/p$, *there exist* $C > 0$ *and* $p' > d$ *depending on* p, d, *such that, for every multiindex* α *of length less than or equal to* q,

$$|\partial_\alpha p_F(x)| \leq C B_{q+1,p'}(F)^{q+1+k_{p,d}} \mathbb{P}(F \in B_2(x))^a. \tag{2.88}$$

In particular, for every l and $|x| > 2$ we have

$$|\partial_\alpha p_F(x)| \leq C B_{q+1,p'}(F)^{q+1+k_{p,d}} \|F\|_{l/a}^l \frac{1}{(|x|-2)^l}. \tag{2.89}$$

Another class of results concern the regularity of conditional expectations. Let $F = (F_1, \ldots, F_d)$ and G be random variables such that the law of F is absolutely continuous with density p_F. We define

$$p_{F,G}(x) = p_F(x)\mathbb{E}(G \mid F = x).$$

This is a measurable function. As a consequence of the general Theorem 1.8.4 we obtain the following result.

Theorem 2.3.2.   (i) *Let $F = (F^1, \ldots, F^d) \in (\mathbb{D}^{2,\infty})^d$ be such that for every $p > 0$ we have $\mathbb{E}(|\det \sigma_F|^{-p}) < \infty$, and let $G \in W^{1,\infty}$. Then, $p_{F,G} \in C(\mathbb{R}^d)$ and it is represented as*

$$p_{F,G}(x) = \sum_{i=1}^{d} \mathbb{E}\big(\partial_i Q_d(F-x)H_i(F,G)\big). \tag{2.90}$$

(ii) *If $F \in (\mathbb{D}^{q+2,\infty})^d$ and $G \in W^{q+1,\infty}$, then $p_{F,G} \in C^q(\mathbb{R}^d)$ and there exist constants C, l such that, for every multiindex α of length less than or equal to q,*

$$\big|\partial_\alpha p_{F,G}(x)\big| \leq C B_{q+1,l}(F)^{q+1+k_{p,d}} \|G\|_{q+1,l}. \tag{2.91}$$

2.4   Comparisons between density functions

We present here some results from [3] allowing to get comparisons between two density functions. In order to show some nontrivial applications of these results, we need to generalize to the concept of "localized probability density functions" that is, densities related to "localized" (probability) measures.

2.4.1   Localized representation formulas for the density

Consider a random variable U taking values on $[0,1]$ and set

$$d\mathbb{P}_U = U d\mathbb{P}.$$

$\mathbb{P}_U$ is a nonnegative measure (but generally not a probability measure) and we set $\mathbb{E}_U$ to be the expectation (integral) with respect to $\mathbb{P}_U$. For $F \in \mathbb{D}^{k,p}$, we define

$$\|F\|_{p,U}^p = \mathbb{E}_U(|F|^p)$$

and

$$\|F\|_{k,p,U}^p = \left(\mathbb{E}_U(|F|^p)\right)^{1/p}$$

$$+ \sum_{r=1}^{k} \sum_{|\alpha|=r} \left(\mathbb{E}\left(\left(\int_{(0,1)^k} |D^\alpha_{s_1,\dots,s_k} F|^2 \, ds_1 \cdots ds_k\right)^{p/2}\right)\right)^{1/p}.$$

In other words, $\|\cdot\|_{p,U}$ and $\|\cdot\|_{k,p,U}$ are the standard Sobolev norms in Malliavin calculus with $\mathbb{P}$ replaced by the localized measure $\mathbb{P}_U$.

For $U \in \mathbb{D}^{q,\infty}$ with $q \in \mathbb{N}^*$, and for $p \in \mathbb{N}$, we set

$$m_{q,p}(U) := 1 + \|D \ln U\|_{q-1,p,U}. \tag{2.92}$$

Equation (2.92) could seem problematic because U may vanish and then $D(\ln U)$ is not well defined. Nevertheless, we make the convention that

$$D(\ln U) = \frac{1}{U} DU \, \mathbf{1}_{\{U \neq 0\}}$$

(in fact this is the quantity we are really concerned with). Since $U > 0$, $\mathbb{P}_U$-a.s. and DU is well defined, the relation $\|\ln U\|_{q,p,U} < \infty$ makes sense.

We give now the integration by parts formula with respect to $\mathbb{P}_U$ (that is, locally) and we study some consequences concerning the regularity of the law, starting from the results in Chapter 1.

Once and for all, in addition to $m_{q,p}(U)$ as in (2.92), we define the following quantities: for $p \geq 1$, $q \in \mathbb{N}$,

$$S_{F,U}(p) = 1 + \|(\det \sigma_F)^{-1}\|_{p,U}, \qquad F \in (\mathbb{D}^{1,\infty})^d$$

$$Q_{F,U}(q,p) = 1 + \|F\|_{q,p,U}, \qquad F \in (\mathbb{D}^{q,\infty})^d \tag{2.93}$$

$$Q_{F,\overline{F},U}(q,p) = 1 + \|F\|_{q,p,U} + \|\overline{F}\|_{q,p,U}, \qquad F \in (\mathbb{D}^{q,\infty})^d,$$

with the convention $S_{F,U}(p) = +\infty$ if the right-hand side is not finite.

Proposition 2.4.1. *Let $\kappa \in \mathbb{N}^*$ and assume that $m_{\kappa,p}(U) < \infty$ for all $p \geq 1$, with $m_{\kappa,p}(U)$ defined as in (2.92). Let $F = (F_1, \dots, F_d) \in (\mathbb{D}^{\kappa+1,\infty})^d$ be such that $S_{F,U}(p) < \infty$ for every $p \in \mathbb{N}$. Let γ_F be the inverse of σ_F on the set $\{U \neq 0\}$. Then, the following localized integration by parts formula holds: for every $f \in C_b^\infty(\mathbb{R}^d)$, $G \in \mathbb{D}^{\kappa,\infty}$, and for every multiindex α of length equal to $q \leq \kappa$, we have*

$$\mathbb{E}_U(\partial_\alpha f(F) \, G) = \mathbb{E}_U\left(f(F) H^q_{\alpha,U}(F,G)\right),$$

where, for $i = 1, \dots, d$,

$$H_{i,U}(F,G) = \sum_{j=1}^{d} \left(G \gamma_F^{ji} L F^j - \langle D(G \gamma_F^{ji}), DF^j \rangle - G \gamma_F^{ji} \langle D \ln U, DF^j \rangle \right)$$

$$= H_i(F,G) - \sum_{j=1}^{d} G \gamma_F^{ji} \langle D \ln U, DF^j \rangle; \tag{2.94}$$

and, for a general multiindex β with $|\beta| = q$,

$$H^q_{\beta,U}(F,G) = H_{\beta_q,U}\big(F, H^{q-1}_{(\beta_1,\dots,\beta_{q-1}),U}(F,G)\big).$$

Moreover, let $l \in \mathbb{N}$ be such that $m_{l+\kappa,p}(U) < \infty$ for all $p \geq 1$, let $F, \overline{F} \in (\mathbb{D}^{l+\kappa+1,\infty})^d$ with $\mathrm{S}_{F,U}(p), \mathrm{S}_{\overline{F},U}(p) < \infty$ for every p, and let $G, \overline{G} \in \mathbb{D}^{l+\kappa,\infty}$. Then, for every $p \geq 1$ there are two universal constants $C > 0$ and $p' > d$ (depending on κ, d) such that, for every multiindex β with $|\beta| = q \leq \kappa$,

$$\|H^q_{\beta,U}(F,G)\|_{l,p,U} \leq C\, B_{l+q,p',U}(F)^q\, \|G\|_{l+q,p',U}, \tag{2.95}$$

$$\|H^q_{\beta,U}(F,G) - H^q_{\beta,U}(\overline{F},\overline{G})\|_{l,p,U} \leq C\, B_{l+q,p',U}(F,\overline{F})^{\frac{q(q+1)}{2}}\, Q_{G,U}(l+q,p',U)$$
$$\times \big(\|F - \overline{F}\|_{l+q+1,p',U} + \|G - \overline{G}\|_{l+q,p',U}\big), \tag{2.96}$$

where

$$B_{l,p,U}(F) = \mathrm{S}_{F,U}(p)^{l+1}\mathrm{Q}_{F,U}(l+1,p)^{2d(l+2)}m_{l,p}(U) \tag{2.97}$$

$$B_{l,p,U}(F,\overline{F}) = \mathrm{S}_{F,U}(p)^{l+1}\mathrm{S}_{\overline{F},U}(p)^{l+1}\mathrm{Q}_{F,\overline{F},U}(l+1,p)^{2d(l+3)}m_{l,p}(U). \tag{2.98}$$

We recall that $\mathrm{S}_{\cdot,U}(p)$, $\mathrm{Q}_{\cdot,U}(l,p)$ and $\mathrm{Q}_{\cdot,\cdot,U}(l,p)$ are defined in (2.93).

Proof. For $|\beta| = 1$, the integration by parts formula immediately follows from the fact that

$$\mathbb{E}_U(\partial_i f(F)G) = \mathbb{E}(\partial_i f(F)GU) = \mathbb{E}(f(F)H_i(F,GU)),$$

so that $H_{i,U}(F,G) = H_i(F,GU)/U$, and this gives the formula for $H_{i,U}(F,G)$. For higher-order integration by parts it suffices to iterate this procedure.

Now, by using the same arguments as in Theorem 2.1.10 for simple functionals, and then passing to the limit as in Theorem 2.2.11, we get that there exists a universal constant C, depending on q and d, such that for every multiindex β of length q we have

$$|H^q_{\beta,U}(F,G)|_l \leq C\, A_{l+q}(F)^q\big(1 + |D\ln U|_{l+q-1}\big)^q |G|_{l+q}$$

and

$$|H^q_{\beta,U}(F,G) - H^q_{\beta,U}(\overline{F},\overline{G})|_l \leq C\, A_{l+q}(F,\overline{F})^{\frac{q(q+1)}{2}}\big(1 + |D\ln U|_{l+q-1}\big)^{\frac{q(q+1)}{2}}$$
$$\times \big(1 + |G|_{l+q} + |\overline{G}|_{l+q}\big)^q$$
$$\times \big(|F - \overline{F}|_{l+q+1} + |L(F - \overline{F})|_{l+q-1} + |G - \overline{G}|_{l+q}\big),$$

where $A_l(F)$ and $A_l(F,\overline{F})$ are defined in (2.37) and (2.38), respectively (as usual, $|\cdot|_0 \equiv |\cdot|$). By using these estimates and by applying the Hölder and Meyer inequalities, one gets (2.95) and (2.96), as done in Corollary 2.2.13. $\square$

Lemma 2.4.2. *Let $\kappa \in \mathbb{N}^*$ and assume that $m_{\kappa,p}(U) < \infty$ for every $p \geq 1$. Let $F \in (\mathbb{D}^{\kappa+1,\infty})^d$ be such that $S_{F,U}(p) < \infty$ for every $p \geq 1$.*

(i) *Let Q_d be the Poisson kernel in $\mathbb{R}^d$. Then, for every $p > d$, there exists a universal constant C depending on d, p such that*

$$\|\nabla Q_d(F - x)\|_{\frac{p}{p-1},U} \leq C\|H_U(F,1)\|_{p,U}^{k_{p,d}}, \tag{2.99}$$

where $k_{p,d} = (d-1)/(1 - d/p)$ and $H_U(F,1)$ denotes the vector in $\mathbb{R}^d$ whose i-th entry is given by $H_{i,U}(F,1)$.

(ii) *Under $\mathbb{P}_U$, the law of F is absolutely continuous and has a density $p_{F,U} \in C^{\kappa-1}(\mathbb{R}^d)$ whose derivatives up to order $\kappa - 1$ may be represented as*

$$\partial_\alpha p_{F,U}(x) = \sum_{i=1}^{d} \mathbb{E}_U\left(\partial_i Q_d(F - x) H_{(i,\alpha),U}^{q+1}(F,1)\right), \tag{2.100}$$

for every multiindex α with $|\alpha| = q \leq \kappa$. Moreover, there exist constants C, a, and $p' > d$ depending on κ, d, such that, for every multiindex α with $|\alpha| = q \leq \kappa - 1$,

$$|\partial_\alpha p_{F,U}(x)| \leq C S_{F,U}(p')^a Q_{F,U}(q+2,p')^a m_U(q+1,p')^a, \tag{2.101}$$

where $m_{k,p}(U)$, $S_{F,U}(p)$ and $Q_{F,U}(q+2,p)$ are given in (2.92) and (2.93), respectively.

Finally, if $V \in \mathbb{D}^{\kappa,\infty}$, then for $|\alpha| = q \leq \kappa - 1$ we have

$$|\partial_\alpha p_{F,UV}(x)| \leq C S_{F,U}(p')^a Q_{F,U}(q+2,p')^a m_U(q+1,p')^a \|V\|_{q+1,p',U}, \tag{2.102}$$

where C, a, and $p' > d$ depend on κ, d.

Proof. (i) This item is actually Theorem 1.8.3 (recall that $\|1\|_{W_{\mu_F}^{1,p}} \leq \|H(F,1)\|_p$) with $\mathbb{P}$ replaced by $\mathbb{P}_U$.

(ii) Equation (2.100) again follows from Theorem 1.8.4 (ii) (with $G = 1$) and, again, by considering $\mathbb{P}_U$ in place of $\mathbb{P}$. As for (2.101), by using the Hölder inequality with $p > d$ and (2.99), we have

$$|\partial_\alpha p_{F,U}(x)| \leq \sum_{i=1}^{d} \|\nabla Q_d(F - x)\|_{\frac{p}{p-1},U} \|H_{(i,\alpha),U}^{q+1}(F,1))\|_{p,U}$$

$$\leq C \sum_{i=1}^{d} \|H_U(F,1)\|_{p,U}^{k_{d,p}} \|H_{(i,\alpha),U}^{q+1}(F,1))\|_{p,U},$$

and the results follow immediately from (2.95).

Take now $V \in \mathbb{D}^{\kappa,\infty}$. For any smooth function f we have

$$\mathbb{E}_{UV}(\partial_i f(F)) = \mathbb{E}_U(V\partial_i f(F)) = \mathbb{E}_U(f(F)H_{i,U}(F,V)).$$

Therefore, by using the results in Theorem 1.8.4, for $|\alpha| = q \leq \kappa - 1$ we get

$$\partial_\alpha p_{UV}(x) = \sum_{i=1}^{d} \mathbb{E}_U\left(\partial_i Q_d(F-x)H_{(i,\alpha),U}^{q+1}(F,V)\right),$$

and proceeding as above we obtain

$$|\partial_\alpha p_{UV}(x)| \leq \sum_{i=1}^{d} \|H_U(F,1)\|_{p,U}^{k_{d,p}} \|H_{(i,\alpha),U}^{q+1}(F,V)\|_{p,U}.$$

Applying again (2.95), the result follows. $\qquad\qquad\qquad\qquad\qquad$ □

2.4.2   The distance between density functions

We compare now the densities of the laws of two random variables under $\mathbb{P}_U$.

Proposition 2.4.3. *Let $q \in \mathbb{N}$ and assume that $m_{q+2,p}(U) < \infty$ for every $p \geq 1$. Let $F, G \in (\mathbb{D}^{q+3,\infty})^d$ be such that*

$$S_{F,G,U}(p) := 1 + \sup_{0 \leq \lambda \leq 1} \|(\det \sigma_{G+\lambda(F-G)})^{-1}\|_{p,U} < \infty, \quad \forall p \in \mathbb{N}. \qquad (2.103)$$

Then, under $\mathbb{P}_U$, the laws of F and G are absolutely continuous with respect to the Lebesgue measure with densities $p_{F,U}$ and $p_{G,U}$, respectively. And, for every multiindex α with $|\alpha| = q$, there exist constants $C, a > 0$ and $p' > d$ depending on d, q such that

$$|\partial_\alpha p_{F,U}(y) - \partial_\alpha p_{G,U}(y)| \leq C\, S_{F,G,U}(p')^a\, Q_{F,G,U}(q+3,p')^a\, m_{q+2,p'}(U)^a \qquad (2.104)$$
$$\times \|F-G\|_{q+2,p',U},$$

where $m_{k,p}(U)$ and $Q_{F,G,U}(k,p)$ are given in (2.92) and (2.93), respectively.

Proof. Throughout this proof, C, p', a will denote constants that can vary from line to line and depending on d, q.

Thanks to Lemma 2.4.2, under $\mathbb{P}_U$ the laws of F and G are both absolutely continuous with respect to the Lebesgue measure and, for every multiindex α with

$|\alpha| = q$, we have

$$\partial_{\alpha} p_{F,U}(y) - \partial_{\alpha} p_{G,U}(y) = \sum_{j=1}^{d} \mathbb{E}_{U}\left(\left(\partial_j Q_d(F-y) - \partial_j Q_d(G-y) \right) H_{(j,\alpha),U}^{q+1}(G,1) \right)$$

$$+ \sum_{j=1}^{d} \mathbb{E}_{U}\left(\partial_j Q_d(F-y) \left(H_{(j,\alpha),U}^{q+1}(F,1) - H_{(j,\alpha),U}^{q+1}(G,1) \right) \right)$$

$$=: \sum_{j=1}^{d} I_j + \sum_{j=1}^{d} J_j.$$

By using (2.99), for $p > d$ we obtain

$$|J_j| \le C \, \|\nabla Q_d(F-y)\|_{\frac{p}{p-1},U} \, \|H_{(j,\alpha),U}^{q+1}(F,1) - H_{(j,\alpha),U}^{q+1}(G,1)\|_{p,U}$$

$$\le C \, \|H_U(F,1)\|_{p,U}^{k_{d,p}} \, \|H_{(j,\alpha),U}^{q+1}(F,1) - H_{(j,\alpha),U}^{q+1}(G,1)\|_{p,U}$$

and, by applying (2.95) and (2.96), there exists $p' > d$ such that

$$|J_j| \le C B_{q+1,p',U}(F,G)^{k_{d,p} + \frac{(q+1)(q+2)}{2}} \, \|F - G\|_{q+2,p',U},$$

with $B_{q+1,p',U}(F,G)$ being defined as in (2.98). By using the quantities $S_{F,G,U}(p)$ and $Q_{F,G,U}(k,p)$, for a suitable $a > 1$ we can write

$$|J_j| \le C S_{F,G,U}(p')^a Q_{F,G,U}(q+2,p')^a m_{q+1,p'}(U)^a \times \|F - G\|_{q+2,p',U}.$$

We now study I_j. For $\lambda \in [0,1]$ we denote $F_{\lambda} = G + \lambda(F - G)$ and we use the Taylor expansion to obtain

$$I_j = \sum_{k=1}^{d} R_{k,j}, \quad \text{with} \quad R_{k,j} = \int_0^1 \mathbb{E}_{U}\left(\partial_k \partial_j Q_d(F_{\lambda}-y) H_{(j,\alpha),U}^{q+1}(G,1)(F-G)_k \right) d\lambda.$$

Let $V_{k,j} = H_{(j,\alpha),U}^{q+1}(G,1)(F-G)_k$. Since for $\lambda \in [0,1]$, $\mathbb{E}_{U}((\det \sigma_{F_{\lambda}})^{-p})) < \infty$ for every p, we can use the integration by parts formula with respect to F_{λ} to get

$$R_{k,j} = \int_0^1 \mathbb{E}_{U}\left(\partial_j Q_d(F_{\lambda}-y) H_{k,U}(F_{\lambda}, V_{k,j}) \right) d\lambda.$$

Therefore, taking $p > d$ and using again (2.99), (2.95) and (2.96), we get

$$|R_{k,j}| \le \int_0^1 \|\partial_j Q_d(F_{\lambda}-y)\|_{\frac{p}{p-1},U} \|H_{k,U}(F_{\lambda}, V_{k,j})\|_{p,U} d\lambda$$

$$\le C \int_0^1 \|H_U(F_{\lambda},1)\|_{p,U}^{k_{d,p}} \|H_{k,U}(F_{\lambda}, V_{k,j})\|_{p,U} d\lambda$$

$$\le C \int_0^1 B_{1,p',U}(F_{\lambda})^{k_{d,p}+1} \|V_{k,j}\|_{1,p'U} \, d\lambda.$$

Now, from (2.103) and (2.97) it follows that

$$B_{1,p,U}(F_\lambda) \leq C S_{F,G,U}^2(p) Q_{F,G,U}^{6d}(2,p) \times m_{1,p}(U).$$

Moreover,

$$\|V_{k,j}\|_{1,p,U} = \|H_{(j,\alpha),U}^{q+1}(G,1)(F-G)_k\|_{1,p,U} \leq \|H_{(j,\alpha),U}^{q+1}(G,1)\|_{1,p',U}\|F-G\|_{1,p',U}$$

$$\leq C\,B_{q+2,p',U}(F)^{q+1}\,\|G\|_{q+2,p',U}\|F-G\|_{1,p',U}$$

and, by using (2.97),

$$B_{q+2,p',U}(F) \leq S_{F,G,U}(p)^{q+3} Q_{F,U}(q+3,p')^{2d(q+4)} \times m_{q+2,p'}(U),$$

where $Q_{F,U}(l,p)$ is defined as in (2.93). So, combining everything, we finally have

$$|I_j| \leq C S_{F,G,U}(p')^a Q_{F,G,U}(q+3,p')^a m_{q+2,p'}(U)^a \|F-G\|_{1,p',U}$$

and the statement follows. $\square$

Example 2.4.4. We give here an example of localizing function giving rise to a localizing random variable U for which $m_{q,p}(U) < \infty$ for every p.

For $a > 0$, set $\psi_a \colon \mathbb{R} \to \mathbb{R}_+$ as

$$\psi_a(x) = 1_{|x|\leq a} + \exp\left(1 - \frac{a^2}{a^2 - (x-a)^2}\right)1_{a<|x|<2a}. \qquad (2.105)$$

Then $\psi_a \in C_b^\infty(\mathbb{R})$, $0 \leq \psi_a \leq 1$. For every $p \geq 1$ we have

$$\sup_x \left|\left(\ln \psi_a(x)\right)'\right|^p \psi_a(x) \leq \frac{4^p}{a^p} \sup_{t\geq 0}\left(t^{2p}e^{1-t}\right) < \infty.$$

For $\Theta_i \in \mathbb{D}^{1,\infty}$ and $a_i > 0$, $i = 1,\ldots,\ell$, we define

$$U = \prod_{i=1}^{\ell} \psi_{a_i}(\Theta_i). \qquad (2.106)$$

Then $U \in \mathbb{D}^{1,\infty}$, $U \in [0,1]$, and $m_{1,p}(U) < \infty$ for every p. In fact, we have

$$|D\ln U|^p U = \left|\sum_{i=1}^{\ell}(\ln\psi_{a_i})'(\Theta_i)D\Theta_i\right|^p \prod_{j=1}^{\ell}\psi_{a_j}(\Theta_j)$$

$$\leq \left(\sum_{i=1}^{\ell}\left|(\ln\psi_{a_i})'(\Theta_i)\right|^2\right)^{p/2}\left(\sum_{i=1}^{\ell}\left|D\Theta_i\right|^2\right)^{p/2}\prod_{j=1}^{\ell}\psi_{a_j}(\Theta_j)$$

$$\leq c_p \sum_{i=1}^{\ell}\left|(\ln\psi_{a_i})'(\Theta_i)\right|^p\psi_{a_i}(\Theta_i)\times|D\Theta|^p$$

$$\leq C_p\sum_{i=1}^{\ell}\frac{1}{a_i^p}\left|D\Theta\right|^p$$

for a suitable $C_p > 0$, so that

$$\mathbb{E}\big(U|D\ln U|^p\big) \le C_p \sum_{i=1}^{\ell} \frac{1}{a_i^p} \times \mathbb{E}(U|D\Theta|^p) \le C_p \sum_{i=1}^{\ell} \frac{1}{a_i^p} \times \|\Theta\|_{1,p,U}^p < \infty.$$

This procedure can be easily generalized to higher-order derivatives: similar arguments give that, for $\Theta \in \mathbb{D}^{q+1,\infty}$,

$$\|D\ln U\|_{q,p,U} \le C_{p,q}\,\|\Theta\|_{q+1,p,U} < \infty. \tag{2.107}$$

Let us propose a further example, that will be used soon.

Example 2.4.5. Let $V = \psi_a(H)$, with ψ_a as in (2.105) and $H \in \mathbb{D}^{q,\infty}$. Consider also a (general) localizing random variable $U \in (0,1)$. Then we have

$$\|V\|_{q,p,U} \le C\,\|H\|_{q,pq,U}^q, \tag{2.108}$$

where C depends on q,p only. In fact, since any derivative of ψ_a is bounded, $|V|_q \le C|H|_q^q$ and (2.108) follows by applying the Hölder inequality.

Using the localizing function in (2.105) and applying Proposition 2.4.3 we get the following result.

Theorem 2.4.6. *Let $q \in \mathbb{N}$. Assume that $m_{q+2,p}(U) < \infty$ for every $p \ge 1$. Let $F, G \in (\mathbb{D}^{q+3,\infty})^d$ be such that $S_{F,U}(p), S_{G,U}(p) < \infty$ for every $p \in \mathbb{N}$. Then, under $\mathbb{P}_U$, the laws of F and G are absolutely continuous with respect to the Lebesgue measure, with densities $p_{F,U}$ and $p_{G,U}$, respectively. Moreover, there exist constants $C, a > 0$, and $p' > d$ depending on q, d such that, for every multiindex α of length q, we have*

$$|\partial_\alpha p_{F,U}(y) - \partial_\alpha p_{G,U}(y)| \le C\,(S_{G,U}(p') \vee S_{F,U}(p'))^a Q_{F,G,U}(q+3,p')^a m_{q+2,p'}(U)^a$$
$$\times \|F - G\|_{q+2,p',U}, \tag{2.109}$$

where $m_{k,p}(U)$ and $Q_{F,G,U}(k,p)$ are as in (2.92) and (2.93), respectively.

Proof. Set $R = F - G$. We use the deterministic estimate (2.28) on the distance between the determinants of two Malliavin covariance matrices: for every $\lambda \in [0,1]$ we can write

$$|\det \sigma_{G+\lambda R} - \det \sigma_G| \le C_d\,|DR|\,(|DG| + |DF|)^{2d-1}$$
$$\le \big(\alpha_d|DR|^2(|DG|^2 + |DF|^2)^{\frac{2d-1}{2}}\big)^{1/2},$$

so that

$$\det \sigma_{G+\lambda R} \ge \det \sigma_G - \alpha_d\big(|DR|^2(|DG|^2 + |DF|^2)^{\frac{2d-1}{2}}\big)^{1/2}. \tag{2.110}$$

For ψ_a as in (2.105), we define

$$V = \psi_{1/8}(H) \quad \text{with} \quad H = |DR|^2 \frac{\left(|DG|^2 + |DF|^2\right)^{\frac{2d-1}{2}}}{(\det \sigma_G)^2},$$

so that

$$V \neq 0 \quad \Longrightarrow \quad \det \sigma_{G+\lambda R} \geq \frac{1}{2} \det \sigma_G. \tag{2.111}$$

Before continuing, let us give the following estimate for the Sobolev norm of H. First, coming back to the notation $|\cdot|_l$ as in (2.15), and using (2.16), we easily get

$$|H|_l \leq C|(\det \sigma_G)^{-1}|_l^2 \left(1 + \left||DF|^2\right|_l + \left||DG|^2\right|_l\right)^d \left||DR|^2\right|_l.$$

By using the estimate concerning the determinant from (2.27) (it is clear that we should pass to the limit for functionals that are not necessarily simple ones) and the straightforward estimate $\left||DF|^2\right|_l \leq C|F|_{l+1}^2$, we have

$$|H|_l \leq C|(\det \sigma_G)^{-1}|^{2(l+1)} \left(1 + |F|_{l+1} + |G|_{l+1}\right)^{6dl} |R|_{l+1}^2.$$

As a consequence, and using the Hölder inequality, we obtain

$$\|H\|_{l,p,U} \leq C S_{G,U}(\bar{p})^{\bar{a}} Q_{F,G,U}(l+1,\bar{p})^{\bar{a}} \|F - G\|_{l+1,\bar{p},U}^2, \tag{2.112}$$

where C, $\bar{p}$, $\bar{a}$ depend on l, d, p.

Now, because of (2.111), we have $S_{F,G,UV}(p) \leq C S_{G,U}(p)$ (C denoting a suitable positive constant which will vary in the following lines). We also have $m_{UV}(k,p) \leq C(m_U(k,p) + m_V(k,p))$. By (2.107) and (2.112), we get

$$m_V(q+2,p) \leq C S_{G,U}(\bar{p})^{\bar{a}} Q_{F,G,U}(q+3,\bar{p})^{\bar{a}}$$

for some $\bar{p}, \bar{a}$, so that $m_{UV}(q+2,p) \leq C S_{G,U}(\bar{p})^{\bar{a}} Q_{F,G,U}(q+3,\bar{p})^{\bar{a}} m_U(q+2,\bar{p})^{\bar{a}}$. Hence, we can apply (2.104) with localization UV and we get

$$|\partial_\alpha p_{F,UV}(y) - \partial_\alpha p_{G,UV}(y)|$$
$$\leq C S_{G,U}(p')^a Q_{F,G,U}(q+3,p')^a m_{q+2,p'}(U)^a \|F - G\|_{q+2,p',U}, \tag{2.113}$$

with $p' > d$, $C > 0$, and $a > 1$ depending on q, d. Further,

$$|\partial_\alpha p_{F,U}(y) - \partial_\alpha p_{G,U}(y)| \leq |\partial_\alpha p_{F,UV}(y) - \partial_\alpha p_{G,UV}(y)|$$
$$+ \left|\partial_\alpha p_{F,U(1-V)}(y)\right| + \left|\partial_\alpha p_{G,UV}(y)\right|,$$

and we have already seen that the first term on the right-hand side behaves as desired. Thus, it suffices to show that the remaining two terms have also the right behavior.

To this purpose, we use (2.102). We have

$$|\partial_\alpha p_{F,U(1-V)}(x)| \le C S_{F,U}(p)^a Q_{F,U}(q+2,p)^a m_U(q+1,p)^a \|1 - V\|_{q+1,p,U}.$$

Now, we can write

$$\|1 - V\|_{q+1,p,U}^p = \mathbb{E}_U(|1 - V|^p) + \|DV\|_{q,p,U}.$$

But $1 - V \ne 0$ implies that $H \ge 1/8$. Moreover, from (2.108), it follows that $\|DV\|_{q,p,U} \le \|V\|_{q+1,p,U} \le C\|H\|_{q+1,p(q+1),U}^{q+1}$. So, we have

$$\|1 - V\|_{q+1,p,U} \le C\big(\mathbb{P}_U(H > 1/8)^{1/p} + \|DV\|_{q,p,U}\big)$$

$$\le C\big(\|H\|_{p,U} + \|H\|_{q+1,p(q+1),U}^{q+1}\big) \le C\|H\|_{q+1,p(q+1),U}^{q+1} \quad (2.114)$$

and, by using (2.112), we get

$$\|1 - V\|_{q+1,p,U} \le C\big(S_{G,U}(\bar{p})^{2(q+2)} Q_{F,G,U}(q+2,\bar{p})^{6d(q+1)} \|F - G\|_{q+2,\bar{p},U}^2\big)^{q+1}.$$

But $\|F - G\|_{q+2,\bar{p},U} \le Q_{F,G,U}(q+2,\bar{p})$, whence

$$|\partial_\alpha p_{F,U(1-V)}(y)|$$

$$\le C\big(S_{F,U}(p') \vee S_{G,U}(p')\big)^a Q_{F,G,U}(p')^a m_U(q+2,p')^a \|F - G\|_{q+2,p',U}$$

for $p' > d$ and suitable constants $C > 0$ and $a > 1$ depending on q, d. Similarly, we get

$$|p_{G,U(1-V)}(y)| \le C S_{G,U}(p')^a Q_{F,G,U}(q+2,p')^a m_U(q+2,p')^a \|F - G\|_{q+2,p',U},$$
$$(2.115)$$

with the same constraints for p', C, a. The statement now follows. $\qquad\square$

2.5 Convergence in total variation for a sequence of Wiener functionals

In the recent years a number of results concerning the weak convergence of functionals on the Wiener space using Malliavin calculus and Stein's method have been obtained by Nourdin, Nualart, Peccati and Poly (see [38, 39] and [40]). In particular, those authors consider functionals living in a finite (and fixed) direct sum of chaoses and prove that, under a very weak non-degeneracy condition, the convergence in distribution of a sequence of such functionals implies the convergence in total variation. Our aim is to obtain similar results for general functionals on the Wiener space which are twice differentiable in Malliavin sense. Other deeper results in this framework are given in Bally and Caramellino [4].

We consider a sequence of d-dimensional functionals F_n, $n \in \mathbb{N}$, which is bounded in $\mathbb{D}^{2,p}$ for every $p \ge 1$. Under a suitable non-degeneracy condition, see (2.118), we prove that the convergence in distribution of such a sequence implies the convergence in the total variation distance. In particular, if the laws of these functionals are absolutely continuous with respect to the Lebesgue measure, this implies the convergence of the densities in L^1. As an application, we use this method in order to study the convergence in the CLT on the Wiener space. Our approach relies heavily on the estimate of the distance between the densities of Wiener functionals proved in Subsection 2.4.2.

We first give a corollary of the results in Section 2.4 (we write it in a non-localized form, even if a localization could be introduced). Let γ_δ be the density of the centered normal law on $\mathbb{R}^d$ of covariance δI. Here, $\delta > 0$ and I is the identity matrix.

Lemma 2.5.1. *Let $q \in \mathbb{N}$ and $F \in (\mathbb{D}^{q+3,\infty})^d$ be such that $\|(\det \sigma_F)^{-1}\|_p < \infty$ for every p. There exist universal positive constants C, l depending on d, q, and there exists $\delta_0 < 1$ depending on d, such that, for every $f \in C^q(\mathbb{R}^d)$ and every multiindex α with $|\alpha| \leq q$ and $\delta < \delta_0$, we have*

$$\left| \mathbb{E}(\partial_\alpha f(F)) - \mathbb{E}(\partial_\alpha (f * \gamma_\delta)(F)) \right| \leq C \|f\|_\infty A_F(q+3, l)^l \sqrt{\delta} \qquad (2.116)$$

with

$$A_F(k, l) = \left(1 + \|(\det \sigma_F)^{-1}\|_l \right)(1 + \|F\|_{k,l}). \qquad (2.117)$$

Proof. During this proof C and l are constants depending on d and q only, which may change from a line to another. We define $F_\delta = F + B_\delta$, where B is an auxiliary d-dimensional Brownian motion independent of W. In the sequel, we consider the Malliavin calculus with respect to (W, B). The Malliavin covariance matrix of F with respect to (W, B) is the same as the one with respect to W (because F does not depend on B). We denote by σ_{F_δ} the Malliavin covariance matrix of F_δ computed with respect to (W, B). We have $\langle \sigma_{F_\delta} \xi, \xi \rangle = \delta |\xi|^2 + \langle \sigma_F \xi, \xi \rangle$. By [16, Lemma 7.29], for every symmetric nonnegative definite matrix Q we have

$$\frac{1}{\det Q} \leq C_1 \int_{\mathbb{R}^d} |\xi|^d e^{-\langle Q\xi, \xi \rangle} d\xi \leq C_2 \frac{1}{\det Q},$$

where C_1 and C_2 are universal constants. Using these two inequalities, we obtain $\det \sigma_{F_\delta} \geq \det \sigma_F / C$ for some universal constant C. So, we obtain $A_{F_\delta}(q, l) \leq C A_F(q, l)$.

We now denote by p_F the density of the law of F and by p_{F_δ} the density of the law of F_δ. Using (2.89) with $p = d + 1$, giving $k_{p,d} = d^2 - 1$, $a = \frac{1}{2d(d+1)}$ and $l = 2d$, we get

$$|\partial_\alpha p_F(x)| \leq C B_{q+1,p'}(F)^{q+d^2} \|F\|_{4d^2(d+1)}^{2d} \frac{1}{(|x| - 2|)^{2d}}.$$

We notice that $B_{q+1,p'}(F)^{q+d^2} \|F\|_{4d^2(d+1)}^{2d} \leq A_F(q+3, l)^l$ for some l depending on q, d, so that

$$|\partial_\alpha p_F(x)| \leq C A_F(q+3, l)^l \frac{1}{(|x| - 2|)^{2d}}.$$

And since $A_{F_\delta}(q+3,l) \leq CA_F(q+3,l)$, we can also write

$$\left|\partial_\alpha p_{F_\delta}(x)\right| \leq CA_F(q+3,l)^l \frac{1}{(|x|-2|)^{2d}}.$$

We use now Theorem 2.4.6: in (2.109), the factor multiplying $\|F-F_\delta\|_{q+2,p',U} = C\delta$ can be bounded from above by $A_F(q+3,l)^l$ for a suitable l, so we get

$$\left|\partial_\alpha p_F(x) - \partial_\alpha p_{F_\delta}(x)\right| \leq C\,A_F(q+3,l)^l\,\delta.$$

We fix now $M>2$. By using these estimates, we obtain

$$\left|\mathbb{E}(\partial_\alpha f(F)) - \mathbb{E}(\partial_\alpha(f*\gamma_\delta)(F))\right|$$
$$= \left|\int f\partial_\alpha p_F dx - \int f\partial_\alpha p_{F_\delta} dx\right|$$
$$\leq \left|\int_{\{|x|\geq M\}} f\partial_\alpha p_F dx\right| + \left|\int_{\{|x|\geq M\}} f\partial_\alpha p_{F_\delta} dx\right| + \left|\int_{\{|x|<M\}} f(\partial_\alpha p - \partial_\alpha p_\delta)dx\right|$$
$$\leq 2C\|f\|_\infty A_F(q+2,l)^l M^{-d} + C\|f\|_\infty A_F(q+3,l)^l M^d \delta$$
$$\leq C\|f\|_\infty A_F(q+3,l)^l(M^{-d} + M^d\delta).$$

So, by optimizing with respect to M, we get (2.116). $\qquad\square$

We now introduce the following distances on the space of probability measures. For $k\geq 0$ and for $f \in C_b^k(\mathbb{R}^d)$ we denote

$$\|f\|_{k,\infty} = \sum_{0\leq|\alpha|\leq k} \sup_{x\in\mathbb{R}^d} |\partial_\alpha f(x)|.$$

Then, for two probability measures μ and ν we define

$$d_k(\mu,\nu) = \sup\left\{\left|\int f d\mu - \int f d\nu\right| : \|f\|_{k,\infty} \leq 1\right\}.$$

For $k=0$ this is the total variation distance, which will be denoted by d_{TV}, and for $k=1$ this is the Fortet–Mourier distance, in symbols d_{FM}. Moreover, for $q\geq 0$ we define

$$d_{-q}(\mu,\nu) = \sup\left\{\sum_{0\leq|\alpha|\leq q}\left|\int \partial_\alpha f d\mu - \int \partial_\alpha f d\nu\right| : \|f\|_\infty \leq 1\right\}.$$

Notice that for $q=0$ we get $d_{-0} = d_{TV}$. We finally recall that if $F_n \to F$ in distribution, then the sequence of the laws μ_{F_n}, $n\in\mathbb{N}$, of F_n converges to the law μ_F of F in the Wasserstein distance d_W, defined by

$$d_W(\mu,\nu) = \sup\left\{\left|\int f d\mu - \int f d\nu\right| : \|\nabla f\|_\infty \leq 1\right\}.$$

The Wasserstein distance d_W is typically defined through Lipschitz functions, but it can be easily seen that the two definitions actually agree. We also notice that $d_W \geq d_1 = d_{FM}$, so the convergence in distribution holds in d_{FM} as well.

We say that a sequence $F_n \in \mathbb{D}^{1,\infty}$, $n \in \mathbb{N}$, is uniformly non-degenerate if for every $p \geq 1$

$$\sup_n \left\| (\det \sigma_{F_n})^{-1} \right\|_p < \infty, \tag{2.118}$$

and we say that a sequence $F_n \in \mathbb{D}^{q,\infty}$, $n \in \mathbb{N}$, is bounded in $\mathbb{D}^{q,\infty}$ if for every $p \geq 1$

$$\sup_n \left\| F_n \right\|_{q,p} < \infty. \tag{2.119}$$

Theorem 2.5.2.   (i) *Let $q \geq 0$ be given. Consider $F, F_n \in \mathbb{D}^{q+3,\infty}$, $n \in \mathbb{N}$, and assume that the sequence $(F_n)_{n \in \mathbb{N}}$ is uniformly non-degenerate and bounded in $\mathbb{D}^{q+3,\infty}$. We denote by μ_F the law of F and by μ_{F_n} the law of F_n. Then for every $k \in \mathbb{N}$,*

$$\lim_n d_k\left(\mu_F, \mu_{F_n}\right) = 0 \quad \Longrightarrow \quad \lim_n d_{-q}\left(\mu_F, \mu_{F_n}\right) = 0. \tag{2.120}$$

Moreover, there exist universal constants C, l depending on q and d, such that, for every $k \in \mathbb{N}$ and for every large n,

$$d_{-q}(\mu_F, \mu_{F_n}) \leq C\left(\sup_n A_{F_n}(q+3,l)^l + A_F(q+3,l)^l \right) d_k(\mu, \mu_n)^{\frac{1}{2k+1}},$$

where $A_{F_n}(k,l)$ and $A_F(k,l)$ are defined in (2.117). This implies that, for every multiindex α with $|\alpha| \leq q$,

$$\int_{\mathbb{R}^d} \left| \partial_\alpha p_{F_n}(x) - \partial_\alpha p_F(x) \right| dx$$
$$\leq C\left(\sup_n A_{F_n}(q+3,l)^l + A_F(q+3,l)^l \right) d_k(\mu, \mu_n)^{\frac{1}{2k+1}}. \tag{2.121}$$

(ii) *Suppose that $F, F_n \in \mathbb{D}^{d+q+3,\infty}$, $n \in \mathbb{N}$, and assume that the sequence $(F_n)_{n \in \mathbb{N}}$ is uniformly non-degenerated and bounded in $\mathbb{D}^{d+q+3,\infty}$. Then for every $k \in \mathbb{N}$, $\lim_n d_k(\mu_F, \mu_{F_n}) = 0$ implies*

$$\left\| \partial_\alpha p_{F_n} - \partial_\alpha p_F \right\|_\infty \leq C\left(\sup_n A_{F_n}(d+q+3,l)^l + A_F(d+q+3,l)^l \right) d_k(\mu, \mu_n)^{\frac{1}{2k+1}} \tag{2.122}$$

for every multiindex α with $|\alpha| \leq q$.

As a consequence of (i) in the above theorem we have that, for $F, F_n \in \mathbb{D}^{q+3,\infty}$, $n \in \mathbb{N}$, such that $(F_n)_{n \in \mathbb{N}}$ is uniformly non-degenerate and bounded in $\mathbb{D}^{q+3,\infty}$, $F_n \to F$ in distribution implies that $d_{-q}(\mu_F, \mu_{F_n}) \to 0$ (because $d_1 \leq d_W$ and convergence in law implies convergence in d_W). In the special case $q = 0$, we have $d_0 = d_{TV}$ so that, for any sequence which is uniformly non-degenerate and bounded in $\mathbb{D}^{3,\infty}$, convergence in law implies convergence in the total variation distance.

Proof of Theorem 2.5.2. During the proof, C stands for a positive constant possibly varying from line to line.

(i) We denote by $\mu * \gamma_\delta$ the probability measure defined by $\int f d(\mu * \gamma_\delta) = \int f * \gamma_\delta d\mu$. We fix a function f such that $\|f\|_\infty \leq 1$ and a multiindex α with $|\alpha| \leq q$, and we write

$$\left| \int \partial_\alpha f d\mu - \int \partial_\alpha f d\mu_n \right| \leq \left| \int \partial_\alpha f d\mu - \int \partial_\alpha f d\mu * \gamma_\delta \right|$$

$$+ \left| \int \partial_\alpha f d\mu_n - \int \partial_\alpha f d\mu_n * \gamma_\delta \right|$$

$$+ \left| \int \partial_\alpha f d\mu * \gamma_\delta - \int \partial_\alpha f d\mu_n * \gamma_\delta \right|.$$

We notice that

$$\int \partial_\alpha f d\mu_n * \gamma_\delta = \int \partial_\alpha (f * \gamma_\delta) d\mu_n = \mathbb{E}\big(\partial_\alpha (f * \gamma_\delta)(F_n)\big)$$

and $\int \partial_\alpha f d\mu_n = \mathbb{E}(\partial_\alpha f(F_n))$ so that, using Lemma 2.5.1, we can find C, l such that, for every small δ,

$$\left| \int \partial_\alpha f d\mu_n - \int \partial_\alpha f d\mu_n * \gamma_\delta \right| = \left| \mathbb{E}(\partial_\alpha f(F_n)) - \mathbb{E}(\partial_\alpha (f * \gamma_\delta)(F_n)) \right|$$

$$\leq C \, \mathrm{A}_{F_n}(q + 3, l) \delta^{1/2}.$$

A similar estimate holds with μ instead of μ_n. We now write

$$\int \partial_\alpha f d\mu_n * \gamma_\delta = \int (\partial_\alpha f) * \gamma_\delta d\mu_n = \int f * \partial_\alpha \gamma_\delta d\mu_n.$$

For each $\delta > 0$ we have

$$\|f * \partial_\alpha \gamma_\delta\|_{k,\infty} \leq \|f\|_\infty \delta^{-k} \leq \delta^{-k}$$

so that

$$\left| \int \partial_\alpha f d\mu * \gamma_\delta - \int \partial_\alpha f d\mu_n * \gamma_\delta \right| = \left| \int f * \partial_\alpha \gamma_\delta d\mu - \int f * \partial_\alpha \gamma_\delta d\mu_n \right|$$

$$\leq \delta^{-k} d_k(\mu, \mu_n).$$

Finally, we obtain

$$\left| \int \partial_\alpha f d\mu - \int \partial_\alpha f d\mu_n \right| \leq C\big(\mathrm{A}_{F_n}(q + 3, l)^l + \mathrm{A}_F(q + 3, l)^l\big)\delta^{1/2} + \delta^{-k} d_k(\mu, \mu_n).$$

Using (2.118) and (2.119) and optimizing with respect to δ, we get

$$d_{-q}(\mu_{F_n}, \mu_F) \leq C\left(\sup_n A_{F_n}(q+3, l)^l + A_F(q+3, l)^l\right) d_k(\mu, \mu_n)^{\frac{1}{2k+1}},$$

where C, l depend on q and d. The proof of (2.120) now follows by passing to the limit.

Assume now that $F_n \to F$ in distribution. Then it converges in the Wasserstein distance d_W and, since $d_W(\mu, \nu) \geq d_1(\mu, \nu)$, we conclude that $F_n \to F$ in d_1 and hence in d_{-q}. We write

$$\int |\partial_\alpha p_{F_n}(x) - \partial_\alpha p_F(x)| \, dx \leq \sup_{\|f\|_\infty \leq 1} \left|\int f(x)\big(\partial_\alpha p_{F_n}(x) - \partial_\alpha p_F(x)\big) dx\right|$$

$$= \sup_{\|f\|_\infty \leq 1} \left|\int \partial_\alpha f(x)\big(p_{F_n}(x) - p_F(x)\big) dx\right|$$

$$= \sup_{\|f\|_\infty \leq 1} \left|\int \partial_\alpha f(x) d\mu_n(x) - \int \partial_\alpha f(x) d\mu(x)\right|$$

$$\leq d_{-q}(\mu_n, \mu),$$

and so (2.121) is also proved.

(ii) To establish (2.122), we write

$$\partial_\alpha p_F(x) = \int_{-\infty}^{x_1} \cdots \int_{-\infty}^{x_d} \partial_1 \cdots \partial_d \partial_\alpha p_F(y) dy$$

and we do the same for $\partial_\alpha p_{F_n}(x)$. This yields

$$\|\partial_\alpha p_F(x) - \partial_\alpha p_{F_n}(x)\|_\infty \leq \int_{\mathbb{R}^d} \big|\partial_1 \cdots \partial_d \partial_\alpha p_F(y) - \partial_1 \cdots \partial_d \partial_\alpha p_{F_n}(y)\big| dy,$$

and the statement follows from (2.121). □

Remark 2.5.3. For $q = 0$, Theorem 2.5.2 can be rewritten in terms of the total variation distance d_{TV}. So, it gives conditions ensuring convergence in the total variation distance.

As a consequence, we can deal with the convergence in the total variation distance and the convergence of the densities in the Central Limit Theorem on the Wiener space. Specifically, let X denote a square integrable d-dimensional functional which is $\mathcal{F}_T$-measurable (here, $\mathcal{F}_T$ denotes the σ-algebra generated by the Brownian motion up to time T), and let $C(X)$ stand for the covariance matrix of X, i.e., $C(X) = (\text{Cov}(X^i, X^j))_{i,j=1,\ldots,d}$, which we assume to be non-degenerate. Let now X_k, $k \in \mathbb{N}$, denote a sequence of independent copies of X and set

$$S_n = \frac{1}{\sqrt{n}} C(X)^{-1/2} \sum_{k=1}^n \big(X_k - \mathbb{E}(X_k)\big). \tag{2.123}$$

The Central Limit Theorem ensures that S_n, $n \in \mathbb{N}$, converges in law to a standard normal law in $\mathbb{R}^d$: for every $f \in C_b(\mathbb{R}^d)$, $\lim_{n \to \infty} \mathbb{E}(f(S_n)) = \mathbb{E}(f(N))$, where N is a standard normal d-dimensional random vector. But convergence in distribution implies convergence in the Fortet–Mourier distance $d_{\mathrm{FM}} = d_1$, so we know that $d_{\mathrm{FM}}(\mu_{S_n}, \mu_N) = d_1(\mu_{S_n}, \mu_N) \to 0$ as $n \to \infty$, where μ_{S_n} denote the law of S_n and μ_N is the standard normal law in $\mathbb{R}^d$. And, thanks to Theorem 2.5.2, we can study the convergence in d_{-q} and, in particular, in the total variation distance d_{TV}.

For $X \in (\mathbb{D}^{1,2})^d$, as usual, we set σ_X to be the associated Malliavin covariance matrix, and we let λ_X denote the smallest eigenvalue of σ_X:

$$\lambda_X = \inf_{|\xi| \leq 1} \langle \sigma_X \xi, \xi \rangle.$$

Theorem 2.5.4. *Let* $X \in (\mathbb{D}^{q+3,\infty})^d$. *Assume that* $\det C(X) > 0$ *and*

$$\mathbb{E}(\lambda_X^{-p}) < \infty \quad \text{for every } p \geq 1.$$

Let μ_{S_n} *denote the law of* S_n *as in* (2.123) *and* μ_N *the standard normal law in* $\mathbb{R}^d$. *Then, there exists a positive constant* C *such that*

$$d_{-q}(\mu_{S_n}, \mu_N) \leq C d_{\mathrm{FM}}(\mu_{S_n}, \mu_N)^{1/3}, \qquad (2.124)$$

where d_{FM} *denotes the Fortet–Mourier distance. As a consequence, for* $q = 0$ *we get*

$$d_{\mathrm{TV}}(\mu_{S_n}, \mu_N) \leq C d_{\mathrm{FM}}(\mu_{S_n}, \mu_N)^{1/3},$$

where d_{TV} *denotes the total variation distance. Moreover, denoting by* p_{S_n} *and* p_N *the density of* S_n *and* N, *respectively, there exists* $C > 0$ *such that*

$$\int_{\mathbb{R}^d} |\partial_\alpha p_{S_n}(x) - \partial_\alpha p_N(x)| \, dx \leq C d_{\mathrm{FM}}(\mu_{S_n}, \mu_N)^{1/3},$$

for every multiindex α *of length* $\leq q$. *Finally, if in addition* $X \in (\mathbb{D}^{d+q+3,\infty})^d$, *then there exists* $C > 0$ *such that for every multiindex* α *with* $|\alpha| \leq q$ *we have*

$$\|\partial_\alpha p_{S_n} - \partial_\alpha p_N\|_\infty \leq C d_{\mathrm{FM}}(\mu_{S_n}, \mu_N)^{1/3}.$$

Remark 2.5.5. In the one-dimensional case, if $\mathbb{E}(\det \sigma_X) > 0$, then $\mathrm{Var}(X) > 0$. Indeed, if $\mathbb{E}(\det \sigma_X) = 0$, then $|DX|^2 = 0$ almost surely, so that X is constant, which is in contradiction with the fact that the variance is non-null. Notice also that in [38] Nualart, Nourdin, and Poly show that, if X is a finite sum of multiple integrals, then the requirement $\mathbb{E}(\det \sigma_X) > 0$ is equivalent to the fact that the law of X is absolutely continuous.

Remark 2.5.6. Theorem 2.5.4 gives a result on the rate of convergence which is not optimal. In fact, Berry–Esseen-type estimates and Edgeworth-type expansions have been intensively studied in the total variation distance, see [15] for a complete overview of the recent research on this topic. The main requirement is that the law of F have an absolutely continuous component. Now, let $F \in (\mathbb{D}^{2,p})^d$ with $p > d$. If $\mathbb{P}(\det \sigma_F > 0) = 1$, σ_F being the Malliavin covariance matrix of F, then the celebrated criterion of Bouleau and Hirsh ensures that the law of F is absolutely continuous (in fact, it suffices that $F \in \mathbb{D}^{1,2}$). But if $\mathbb{P}(\det \sigma_F > 0) < 1$ this criterion no longer works (and examples with the law of F not being absolutely continuous can be easily produced). In [4] we proved that if $\mathbb{P}(\det \sigma_F > 0) > 0$, then the law of F is "locally lower bounded by the Lebesge measure", and this implies that the law of F has an absolutely continuous component (see [7] for details). Finally, the coefficients in the CLT expansion in the total variation distance usually involve the inverse of the Fourier transform (see [15]), but under the "locally lower bounded by the Lebesge measure" property we have recently given in [7] an explicit formula for such coefficients in terms of a linear combination of Hermite polynomials.

Proof of Theorem 2.5.4. Notice first that we may assume, without loss of generality, that $C(X)$ is the identity matrix and $\mathbb{E}(X) = 0$. If not, we prove the result for $\overline{X} = C^{-1/2}(X)(X - \mathbb{E}(X))$ and make a change of variable at the end.

In order to use Theorem 2.5.2 with $F_n = S_n$ and $F = N$, we need that all the random variables involved be defined on the same probability space (this is just a matter of notation). So, we take the independent copies X_k, $k \in \mathbb{N}$, of X in the following way. Suppose that X is $\mathcal{F}_T$-measurable for some $T > 0$, so it is a functional of W_t, $t \leq T$. Then we define X_k to be the same functional as X, but with respect to $W_t - W_{kT}$, $kT \leq t \leq (k+1)T$. Hence, $D_s X_k = 0$ if $s \notin [kT, (k+1)T]$ and the law of $(D_s X_k)_{s \in [kT,(k+1)T]}$ is the same as the law of $(D_s X)_{s \in [0,T]}$. A similar assertion holds for the Malliavin derivatives of higher order. In particular, we obtain

$$|S_n|_l^2 = \frac{1}{n} \sum_{k=1}^n |X_k|_l^2.$$

We conclude that, for every $p > 1$,

$$\|S_n\|_{l,p} \leq \|X\|_{l,p}$$

so that $S_n \in (\mathbb{D}^{q+3,\infty})^d$. We take $N = W_T/\sqrt{T}$, so that $N \in (\mathbb{D}^{q+3,\infty})^d$ and $\|(\det \sigma_N)^{-1}\|_p < \infty$ for every p. Let us now discuss condition (2.118) with $F_n = S_n$. We consider the Malliavin covariance matrix of S_n. We have $\langle DX_k^i, DX_p^j \rangle = 0$ if $k \neq p$, so

$$\sigma_{S_n}^{i,j} = \langle DS_n^i, DS_n^j \rangle = \frac{1}{n} \sum_{k=1}^n \langle DX_k^i, DX_k^j \rangle = \frac{1}{n} \sum_{k=1}^n \sigma_{X_k}^{i,j}.$$

Using the above formula, we obtain

$$\lambda_{S_n} \geq \frac{1}{n} \sum_{k=1}^n \lambda_{X_k},$$

where $\lambda_{S_n} \geq 0$ and $\lambda_{X_k} \geq 0$ denote the smallest eigenvalue of σ_{S_n} and σ_{X_k}, respectively. By using the Jensen and Hölder inequalities, for $p > 1$ we obtain

$$\left(\frac{1}{\lambda_{S_n}}\right)^p \leq \frac{1}{n}\sum_{k=1}^{n}\left(\frac{1}{\lambda_{X_k}}\right)^p,$$

so that

$$\mathbb{E}(\lambda_{S_n}^{-p}) \leq \mathbb{E}(\lambda_X^{-p})$$

and (2.118) does indeed hold. Now, (2.124) immediately follows by applying Theorem 2.5.2 with $k = 1$. And since $d_{-0} = d_{TV}$, we also get the estimate in the total variation distance under the same hypothesis with $q = 0$. All the remaining statements are immediate consequences of Theorem 2.5.2. $\qquad\square$

Chapter 3

Regularity of probability laws by using an interpolation method

One of the outstanding applications of Malliavin calculus is the criterion of regularity of the law of a functional on the Wiener space (presented in Section 2.3). The functional involved in such a criterion has to be regular in Malliavin sense, i.e., it has to belong to the domain of the differential operators in this calculus. As long as solutions of stochastic equations are concerned, this amounts to regularity properties of the coefficients of the equation. Recently, Fournier and Printems [26] proposed another approach which permits to prove the absolute continuity of the law of the solution of some equations with Hölder coefficients —and such functionals do not fit in the framework of Malliavin calculus (are not differentiable in Malliavin sense). This approach is based on the analysis of the Fourier transform of the law of the random variable at hand. It has already been used in [8, 9, 10, 21, 26]. Recently, Debussche and Romito [20] proposed a variant of the above mentioned approach based on a relative compactness criterion in Besov spaces. It has been used by Fournier [25] in order to study the regularity of the solution of the three-dimensional Boltzmann equation, and by Debussche and Fournier [19] for jump type equations.

The aim of this chapter is to give an alternative approach (following [5]) based on expansions in Hermite series —developed in Appendix A— and on some strong estimates concerning kernels built by mixing Hermite functions —studied in Appendix C. We give several criteria concerning absolute continuity and regularity of the law: Theorems 3.2.3, 3.2.4 and 3.3.2. And in Section 3.4 we use them in the framework of path dependent diffusion processes and stochastic heat equation. Moreover, we finally notice that our approach fits in the theory of interpolation spaces, see Appendix B for general results and references.

3.1   Notations

We work on $\mathbb{R}^d$ and we denote by $\mathcal{M}$ the set of the finite signed measures on $\mathbb{R}^d$ with the Borel σ-algebra. Moreover, $\mathcal{M}_a \subset \mathcal{M}$ is the set of measures which are absolutely continuous with respect to the Lebesgue measure. For $\mu \in \mathcal{M}_a$ we denote by p_μ the density of μ with respect to the Lebesgue measure. And for a measure $\mu \in \mathcal{M}$, we denote by L^p_μ the space of measurable functions $f : \mathbb{R}^d \to \mathbb{R}$ such that $\int |f|^p \, d\mu < \infty$. For $f \in L^1_\mu$ we denote $f\mu \in \mathcal{M}$ the measure $(f\mu)(A) =$

$\int_A f d\mu$. For a bounded function $\phi \colon \mathbb{R}^d \to \mathbb{R}$ we denote $\mu * \phi$ the measure defined by $\int f d\mu * \phi = \int f * \phi d\mu = \int \int \phi(x - y) f(y) dy d\mu(x)$. Then, $\mu * \phi \in \mathcal{M}_a$ and $p_{\mu*\phi}(x) = \int \phi(x - y) d\mu(y)$.

From now on, we consider multiindices which are slightly different from the ones we have worked with. One could pass from the old to the new definition but, for the sake of clarity, in the context we are going to introduce it, it is more convenient to deal with the new one. So, a multiindex α is now of the form $\alpha = (\alpha_1, \ldots, \alpha_d) \in \mathbb{N}^d$. Here, we put $|\alpha| = \sum_{i=1}^d \alpha_i$. For a multiindex α with $|\alpha| = k$ we denote by ∂_α the corresponding derivative $\partial_{x_1}^{\alpha_1} \cdots \partial_{x_d}^{\alpha_d}$, with the convention that $\partial_{x_i}^{\alpha_i} f = f$ if $\alpha_i = 0$. In particular, we allow α to be the null multiindex and in this case one has $\partial_\alpha f = f$. In the following, we set $\mathbb{N}_* = \mathbb{N} \setminus \{0\}$.

We denote $\|f\|_p = (\int |f(x)|^p \, dx)^{1/p}$, $p \geq 1$ and $\|f\|_\infty = \sup_{x \in \mathbb{R}^d} |f(x)|$. Then $L^p = \{f : \|f\|_p < \infty\}$ are the standard L^p spaces with respect to the Lebesgue measure. For $f \in C^k(\mathbb{R}^d)$ we define the norms

$$\|f\|_{k,p} = \sum_{0 \leq |\alpha| \leq k} \|\partial_\alpha f\|_p \quad \text{and} \quad \|f\|_{k,\infty} = \sum_{0 \leq |\alpha| \leq k} \|\partial_\alpha f\|_\infty, \qquad (3.1)$$

and we denote

$$W^{k,p} = \{f : \|f\|_{k,p} < \infty\} \quad \text{and} \quad W^{k,\infty} = \{f : \|f\|_{k,\infty} < \infty\}.$$

For $l \in \mathbb{N}$ and a multiindex α, we denote $f_{\alpha,l}(x) = (1 + |x|)^l \partial_\alpha f(x)$. Then, we consider the weighted norms

$$\|f\|_{k,l,p} = \sum_{0 \leq |\alpha| \leq k} \|f_{\alpha,l}\|_p \quad \text{and} \quad W^{k,l,p} = \{f : \|f\|_{k,l,p} < \infty\}. \qquad (3.2)$$

We stress that in $\| \cdot \|_{k,l,p}$ the first index k is related to the order of the derivatives which are involved, while the second index l is connected to the power of the polynomial multiplying the function and its derivatives up to order k.

3.2   Criterion for the regularity of a probability law

For two measures $\mu, \nu \in \mathcal{M}$ and for $k \in \mathbb{N}$, we consider the distance $d_k(\mu, \nu)$ introduced in Section 2.5, that is

$$d_k(\mu, \nu) = \sup \left\{ \left| \int \phi d\mu - \int \phi d\nu \right| : \phi \in C^\infty(\mathbb{R}^d), \|\phi\|_{k,\infty} \leq 1 \right\}. \qquad (3.3)$$

We recall that $d_0 = d_{TV}$ (total variation distance) and $d_1 = d_{FM}$ (Fortet–Mourier distance). We also recall that $d_1(\mu, \nu) \leq d_W(\mu, \nu)$, where d_W is the (more popular) Wasserstein distance

$$d_W(\mu, \nu) = \sup\{|\int \phi d\mu - \int \phi d\nu| : \phi \in C^1(\mathbb{R}^d), \|\nabla \phi\|_\infty \leq 1\}.$$

We finally recall that the Wasserstein distance is relevant from a probabilistic point of view because it characterizes the convergence in law of probability measures. The distances d_k with $k \geq 2$ are less often used. We mention, however, that in Approximation Theory (for diffusion processes, for example, see [37, 46]) such distances are used in an implicit way: indeed, those authors study the rate of convergence of certain schemes, but they are able to obtain their estimates for test functions $f \in C^k$ with k sufficiently large (so d_k comes on).

Let $q, k \in \mathbb{N}$, $m \in \mathbb{N}_*$, and $p > 1$. We denote by p_* the conjugate of p. For $\mu \in \mathcal{M}$ and for a sequence $\mu_n \in \mathcal{M}_a$, $n \in \mathbb{N}$, we define

$$\pi_{q,k,m,p}(\mu, (\mu_n)_n) = \sum_{n=0}^{\infty} 2^{n(q+k+d/p_*)} d_k(\mu, \mu_n) + \sum_{n=0}^{\infty} \frac{1}{2^{2nm}} \|p_{\mu_n}\|_{2m+q,2m,p}. \quad (3.4)$$

We also define

$$\rho_{q,k,m,p}(\mu) = \inf \pi_{q,k,m,p}(\mu, (\mu_n)_n) \quad (3.5)$$

where the infimum is taken over all the sequences of measures μ_n, $n \in \mathbb{N}$, which are absolutely continuous. It is easy to check that $\rho_{q,k,m,p}$ is a norm on the space $\mathcal{S}_{q,k,m,p}$ defined by

$$\mathcal{S}_{q,k,m,p} = \{\mu \in \mathcal{M} : \rho_{q,k,m,p}(\mu) < \infty\}. \quad (3.6)$$

We prove in Appendix B that

$$\mathcal{S}_{q,k,m,p} = (X, Y)_\gamma,$$

with

$$X = W^{q+2m,2m,p}, \quad Y = W_*^{k,\infty}, \quad \gamma = \frac{q+k+d/p_*}{2m},$$

and where $(X, Y)_\gamma$ denotes the interpolation space of order γ between the Banach spaces X and Y. The space $W_*^{k,\infty}$ is the dual of $W^{k,\infty}$ and we notice that $d_k(\mu, \nu) = \|\mu - \nu\|_{W_*^{k,\infty}}$.

The following proposition is the key estimate here. We prove it in Appendix A.

Proposition 3.2.1. *Let $q, k \in \mathbb{N}$, $m \in \mathbb{N}_*$, and $p > 1$. There exists a universal constant C (depending on q, k, m, d and p) such that, for every $f \in C^{2m+q}(\mathbb{R}^d)$, one has*

$$\|f\|_{q,p} \leq C\rho_{q,k,m,p}(\mu), \quad (3.7)$$

where $\mu(x) = f(x)dx$.

We state now our main theorem:

Theorem 3.2.2. *Let $q, k \in \mathbb{N}, m \in \mathbb{N}_*$ and let $p > 1$.*

(i) *Take $q = 0$. Then, $\mathcal{S}_{0,k,m,p} \subset L^p$ in the sense that if $\mu \in \mathcal{S}_{0,k,m,p}$, then μ is absolutely continuous and the density p_μ belongs to L^p. Moreover, there exists a universal constant C such that*

$$\|p_\mu\|_p \leq C\rho_{0,k,m,p}(\mu).$$

(ii) *Take $q \geq 1$. Then*

$$\mathcal{S}_{q,k,m,p} \subset W^{q,p} \quad and \quad \|p_\mu\|_{q,p} \leq C\rho_{q,k,m,p}(\mu), \quad \mu \in \mathcal{S}_{q,k,m,p}.$$

(iii) *If $\frac{p-1}{p} < \frac{2m+q+k}{d(2m-1)}$, then*

$$W^{q+1,2m,p} \subset \mathcal{S}_{q,k,m,p} \subset W^{q,p}$$

and there exists a constant C such that

$$\frac{1}{C}\|p_\mu\|_{q,m,p} \leq \rho_{q,k,m,p}(\mu) \leq C\|p_\mu\|_{q+1,m,p}. \tag{3.8}$$

Proof. Step 1 (Regularization). For $\delta \in (0,1)$ we consider the density γ_δ of the centred Gaussian probability measure with variance δ. Moreover, we consider a truncation function $\Phi_\delta \in C^\infty$ such that $1_{B_{\delta^{-1}}(0)} \leq \Phi_\delta \leq 1_{B_{1+\delta^{-1}}(0)}$ and its derivatives of all orders are bounded uniformly with respect to δ. Then we define

$$T_\delta : C^\infty \longrightarrow C^\infty, \quad T_\delta f = (\Phi_\delta f) * \gamma_\delta, \tag{3.9}$$
$$\widetilde{T}_\delta : C^\infty \longrightarrow C_c^\infty, \quad \widetilde{T}_\delta f = \Phi_\delta(f * \gamma_\delta).$$

Moreover, for a measure $\mu \in \mathcal{M}$, we define $T_\delta^* \mu$ by

$$\langle T_\delta^* \mu, \phi \rangle = \langle \mu, T_\delta \phi \rangle.$$

Then $T_\delta^* \mu$ is an absolutely continuous measure with density $p_{T_\delta^* \mu} \in C_c^\infty$ given by

$$p_{T_\delta^* \mu}(y) = \Phi_\delta(y) \int \gamma_\delta(x - y) d\mu(x).$$

Step 2. Let us prove that for every $\mu \in \mathcal{M}$ and $\mu_n(dx) = f_n(x)dx$, $n \in \mathbb{N}$, we have

$$\pi_{q,k,m,p}\left(T_\delta^* \mu, (T_\delta^* \mu_n)_n\right) \leq C\pi_{q,k,m,p}(\mu, (\mu_n)_n). \tag{3.10}$$

Since $\|T_\delta \phi\|_{k,\infty} \leq C\|\phi\|_{k,\infty}$, we have $d_k(T_\delta^* \mu, T_\delta^* \mu_n) \leq Cd_k(\mu, \mu_n)$.

For $\mu_n(dx) = f_n(x)dx$ we have $p_{T_\delta^* \mu_n}(y) = \widetilde{T}_\delta f_n$. Let us now check that

$$\|\widetilde{T}_\delta f_n\|_{2m+q,2m,p} \leq C\|f_n\|_{2m+q,2m,p}. \tag{3.11}$$

For a measurable function $g\colon \mathbb{R}^d \to \mathbb{R}$ and $\lambda \geq 0$, we put $g_\lambda(x) = (1 + |x|)^\lambda g(x)$. Since $(1 + |x|)^\lambda \leq (1 + |y|)^\lambda (1 + |x - y|)^\lambda$, it follows that

$$(\widetilde{T}_\delta g)_\lambda(x) = (1 + |x|)^\lambda \Phi_\delta(x) \int_{\mathbb{R}^d} \gamma_\delta(y) g(x - y) dy$$

$$\leq \int_{\mathbb{R}^d} \gamma_{\delta,\lambda}(y) g_\lambda(x - y) dy = \gamma_{\delta,\lambda} * g_\lambda(x).$$

Then, by (3.47), $\|(\widetilde{T}_\delta g)_\lambda\|_p \leq \|\gamma_{\delta,\lambda} * g_\lambda\|_p \leq C \|\gamma_{\delta,\lambda}\|_1 \|g_\lambda\|_p \leq C \|g_\lambda\|_p$. Using this inequality (with $\lambda = 2m$) for $g = \partial_\alpha f_n$ we obtain (3.11). Hence, (3.10) follows.

Step 3. Let $\mu \in S_{q,k,m,p}$ so that $\rho_{q,k,m,p}(\mu) < \infty$. Using (3.7), we have $\|T_\delta^* \mu\|_{q,p} \leq \rho_{q,k,m,p}(T_\delta^* \mu)$ and, moreover, using (3.10),

$$\sup_{\delta \in (0,1)} \|T_\delta^* \mu\|_{q,p} \leq C \sup_{\delta \in (0,1)} \rho_{q,k,m,p}(T_\delta^* \mu) \leq \rho_{q,k,m,p}(\mu) < \infty. \tag{3.12}$$

So, the family $T_\delta^* \mu, \delta \in (0,1)$ is bounded in $W^{q,p}$, which is a reflexive space. So it is weakly relatively compact. Consequently, we may find a sequence $\delta_n \to 0$ such that $T_{\delta_n}^* \mu \to f \in W^{q,p}$ weakly. It is easy to check that $T_{\delta_n}^* \mu \to \mu$ weakly, so $\mu(dx) = f(x) dx$ and $f \in W^{q,p}$. As a consequence of (3.12), $\|\mu\|_{q,p} \leq C\rho_{q,k,m,p}(\mu)$.

Step 4. Let us prove (iii). Let $f \in W^{q+1,2m,p}$ and $\mu_f(dx) = f(x) dx$. We have to show that $\rho_{q,k,m,p}(\mu_f) < \infty$. We consider a superkernel ϕ (see (3.57)) and define $f_\delta = f * \phi_\delta$. We take $\delta_n = 2^{-\theta n}$ with θ to be chosen in a moment. Using (3.58), we obtain $d_k(\mu_f, \mu_{f_{\delta_n}}) \leq C \|f\|_{q+1,m,p} \delta_n^{q+k+1}$ and, using (3.59), we obtain $\|f_{\delta_n}\|_{2m+q,2m,p} \leq C \|f\|_{q+1,m,p} \delta_n^{2m-1}$. Then we write, with $C_f = C \|f\|_{q+1,m,p}$,

$$\pi_{q,k,m,p}(\mu_f, \mu_{f_{\delta_n}}) = \sum_{n=0}^{\infty} 2^{n(q+k+d/p_*)} d_k(\mu_f, \mu_{f_{\delta_n}}) + \sum_{n=0}^{\infty} \frac{1}{2^{2nm}} \|f_{\delta_n}\|_{2m+q,2m,p}$$

$$\leq C_f \sum_{n=0}^{\infty} 2^{n(q+k+d/p_* - \theta(q+k+1))} + C_f \sum_{n=0}^{\infty} \frac{1}{2^{n(2m-\theta(2m-1))}}.$$

In order to obtain the convergence of the above series we need to choose θ such that

$$\frac{q + k + d/p_*}{q + k + 1} < \theta < \frac{2m}{2m - 1},$$

and this is possible under our restriction on p. □

We give now a criterion in order to check that $\mu \in S_{q,k,m,p}$. For $C, r > 0$ we denote by $\Lambda(C, r)$ the family of the continuous functions λ such that

(i) $\lambda\colon (0,1) \to \mathbb{R}_+$ is strictly decreasing,

(ii) $\lim_{\delta \to 0} \lambda(\delta) = \infty$,   (3.13)

(iii) $\lambda(\delta) \leq C\delta^{-r}$.

Theorem 3.2.3. *Let $q, k \in \mathbb{N}$, $m \in \mathbb{N}_*$, $p > 1$, and set*

$$\eta > \frac{q + k + d/p_*}{2m}. \tag{3.14}$$

We consider a nonnegative finite measure μ and a family of finite nonnegative and absolutely continuous measures $\mu_\delta(dx) = f_\delta(x)dx$, $\delta > 0$, $f_\delta \in W^{2m+q,2m,p}$. We assume that there exist $C, C', r > 0$, and $\lambda_{q,m,p} \in \Lambda(C,r)$ such that

$$\|f_\delta\|_{2m+q,2m,p} \le \lambda_{q,m,p}(\delta) \quad and \quad \lambda_{q,m,p}(\delta)^\eta d_k(\mu, \mu_\delta) \le C'. \tag{3.15}$$

Then, $\mu(dx) = f(x)dx$ with $f \in W^{q,p}$.

Proof. Fix $\varepsilon_0 > 0$. From (3.13), $\lambda_{q,m,p}$ is a bijection from $(0,1)$ to $(\lambda_{q,m,p}(1), \infty)$. So, if $n^{-(1+\varepsilon_0)}2^{2mn} \ge \lambda_{q,m,p}(1)$ we may take $\delta_n = \lambda_{q,m,p}^{-1}(n^{-(1+\varepsilon_0)}2^{2mn})$. With this choice we have $\|f_{\delta_n}\|_{2m+q,2m,p} \le \lambda_{q,m,p}(\delta_n) = n^{-(1+\varepsilon_0)}2^{2mn}$, so that

$$\sum_{n=1}^{\infty} \frac{1}{2^{2nm}} \|f_{\delta_n}\|_{2m+q,2m,p} < \infty.$$

By the choice of η we also have

$$2^{n(q+k+d/p_*)}d_k(\mu, \mu_{\delta_n}) \le 2^{n(q+k+d/p_*)} \times \frac{C'}{\lambda_{q,m,p}(\delta_n)^\eta} \le 2^{n(q+k+d/p_*)} \times \frac{n^{\eta(1+\varepsilon_0)}}{2^{n\varepsilon_1}},$$

where $\varepsilon_1 = 2m\eta - (q + k + d/p_*) > 0$. Hence,

$$\sum_{n=1}^{\infty} 2^{n(q+k+d/p_*)}d_k(\mu, \mu_{\delta_n}) < \infty, \tag{3.16}$$

and consequently $\pi_{q,k,m,p}(\mu, (\mu_n)) < \infty$. $\square$

Now we go further and notice that if (3.14) holds, then the convergence of the series in (3.16) is very fast. This allows us to obtain some more regularity.

Theorem 3.2.4. *Let $q, k \in \mathbb{N}$, $m \in \mathbb{N}_*$, and $p > 1$. We consider a nonnegative finite measure μ and a family of finite nonnegative measures $\mu_\delta(dx) = f_\delta(x)dx$, $\delta > 0$, and $f_\delta \in W^{2m+q+1,2m,p}$, such that*

$$\int (1 + |x|)d\mu(x) + \sup_{\delta > 0} \int (1 + |x|)d\mu_\delta(x) < \infty. \tag{3.17}$$

Suppose that there exist $C, C', r > 0$, and $\lambda_{q+1,m,p} \in \Lambda(C,r)$ such that

$$\|f_{\delta'}\|_{2m+q+1,2m,p} \le \lambda_{q+1,m,p}(\delta)$$

and

$$\lambda_{q+1,m,p}(\delta)^\eta d_k(\mu, \mu_\delta) \le C', \quad with \quad \eta > \frac{q + k + d/p_*}{2m}. \tag{3.18}$$

We denote

$$s_\eta(q, k, m, p) = \frac{2mn - (q + k + d/p_*)}{2m\eta} \wedge \frac{\eta}{1 + \eta}. \qquad (3.19)$$

Then, for every multiindex α with $|\alpha| = q$ and every $s < s_\eta(q, k, m, p)$, we have $\partial_\alpha f \in \mathcal{B}^{s,p}$, where $\mathcal{B}^{s,p}$ is the Besov space of index s.

Proof. The fact that (3.18) implies $\mu(dx) = f(x)dx$ with $f \in W^{q,p}$ is an immediate consequence of Theorem 3.2.3. We prove now the regularity property: $g := \partial_\alpha f \in \mathcal{B}^{s,p}$ for $|\alpha| = q$ and $s < s_\eta(q, k, m)$. In order to do this we will use Lemma 3.6.1, so we have to check (3.56).

Step 1. We begin with (3.56)(i). The reasoning is analogous with the one in the proof of Theorem 3.2.3, but we will use the first inequality in (3.8) with q replaced by $q + 1$ and k replaced by $k - 1$. So we define $\delta_n = \lambda_{q+1,m,p}^{-1}(n^{-2}2^{2mn})$ and, from (3.13)(iii), we have $\delta_n \le C^{1/r}2^{-\theta n}$ for $\theta < m/2r$. We obtain

$$\|g * \partial_i \phi_\varepsilon\|_p = \|\partial_i \partial_\alpha (f * \phi_\varepsilon)\|_p \le \|f * \phi_\varepsilon\|_{q+1,p} \le \rho_{q+1,k-1,m,p}(\mu * \phi_\varepsilon)$$

$$\le \sum_{n=1}^{\infty} 2^{n(q+k+d/p_*)} d_{k-1}(\mu * \phi_\varepsilon, \mu_{\delta_n} * \phi_\varepsilon)$$

$$+ \sum_{n=1}^{\infty} 2^{-2mn} \|f_{\delta_n} * \phi_\varepsilon\|_{2m+q+1,2m,p}.$$

By the choice of δ_n,

$$\|f_{\delta_n} * \phi_\varepsilon\|_{2m+q+1,2m,p} \le \|f_{\delta_n}\|_{2m+q+1,2m,p} \le \lambda_{q+1,m,p}(\delta_n) \le \frac{1}{n^2}2^{nm}$$

so the second series is convergent.

We now estimate the first sum. Since $\mu * \phi_\varepsilon(\mathbb{R}^d) = \mu(\mathbb{R}^d)$ and $\mu_{\delta_n} * \phi_\varepsilon(\mathbb{R}^d) = \mu_{\delta_n}(\mathbb{R}^d)$, hypothesis (3.17) implies that $\sup_{\varepsilon,n} d_0(\mu * \phi_\varepsilon, \mu_{\delta_n} * \phi_\varepsilon) \le C < \infty$. It is also easy to check that

$$d_{k-1}(\mu * \phi_\varepsilon, \mu_{\delta_n} * \phi_\varepsilon) \le \frac{1}{\varepsilon}d_k(\mu, \mu_{\delta_n})\left(\int |x| \, d\mu(x) + \int |x| \, d\mu_{\delta_n}(x)\right)$$

so that, using (3.18) and the choice of δ_n, we obtain

$$2^{n(q+k+d/p_*)} d_{k-1}(\mu * \phi_\varepsilon, \mu_{\delta_n} * \phi_\varepsilon) \le \frac{C}{\varepsilon}2^{n(q+1+d/p_*)} d_k(\mu, \mu_{\delta_n})$$

$$\le \frac{C}{\varepsilon}2^{n(q+1+d/p_*)} \frac{1}{\lambda_{q+1,m,p}(\delta_n)^\eta}$$

$$\le \frac{Cn^{2\eta}}{\varepsilon}2^{n(q+1+d/p_*-2m\eta)}.$$

Now, fix $\varepsilon > 0$, take an $n_\varepsilon \in \mathbb{N}$ (to be chosen in the sequel), and write

$$\sum_{n=1}^{\infty} 2^{n(q+k+d/p_*)} d_{k-1}(\mu * \phi_\varepsilon, \mu_{\delta_n} * \phi_\varepsilon) \leq C \sum_{n=1}^{n_\varepsilon} 2^{n(q+k+d/p_*)}$$

$$+ \frac{C}{\varepsilon} \sum_{n=n_\varepsilon+1}^{\infty} n^{2\eta} 2^{n(q+k+d/p_*-2\eta m)}.$$

We take $a > 0$ and we upper bound the above series by

$$2^{n_\varepsilon(q+k+d/p_*+a)} + \frac{C}{\varepsilon} 2^{n_\varepsilon(q+k+d/p_*+a-2\eta m)}.$$

In order to optimize, we take n_ε such that $2^{2m\eta n_\varepsilon} = \frac{1}{\varepsilon}$. With this choice we obtain

$$2^{n_\varepsilon(q+k+d/p_*+a)} \leq C\varepsilon^{-\frac{q+k+d/p_*+a}{2m\eta}}.$$

We conclude that

$$\|g * \partial_i \phi_\varepsilon\|_p \leq C\varepsilon^{-\frac{q+k+d/p_*+a}{2m\eta}},$$

which means that (3.56)(i) holds for $s < 1 - (q+k+d/p_*)/(2m\eta)$.

Step 2. Let us check now (3.56)(ii). Take $u \in (0,1)$ (to be chosen in a moment) and define

$$\delta_{n,\varepsilon} = \lambda_{q+1,m,p}^{-1}(n^{-2}2^{2mn} \times \varepsilon^{-(1-u)}).$$

Then we proceed as in the previous step:

$$\left\|\partial_i(g * \phi_\varepsilon^i)\right\|_p \leq \rho_{q+1,k-1,m,p}(\mu * \phi_\varepsilon^i)$$

$$\leq \sum_{n=1}^{\infty} 2^{n(q+k+d/p_*)} d_{k-1}(\mu * \phi_\varepsilon^i, \mu_{\delta_{n,\varepsilon}} * \phi_\varepsilon^i)$$

$$+ \sum_{n=1}^{\infty} 2^{-2mn} \left\|f_{\delta_{n,\varepsilon}} * \phi_\varepsilon^i\right\|_{2m+q+1,2m,p}.$$

It is easy to check that $\left\|h * \phi_\varepsilon^i\right\|_p \leq \varepsilon \|h\|_p$ for every $h \in L^p$, so that, by our choice of $\delta_{n,\varepsilon}$, we obtain

$$\left\|f_{\delta_{n,\varepsilon}} * \phi_\varepsilon^i\right\|_{2m+q+1,2m,p} \leq \varepsilon \left\|f_{\delta_{n,\varepsilon}}\right\|_{2m+q+1,2m,p} \leq \varepsilon \times \frac{2^{2mn}}{n^2} \times \varepsilon^{-(1-u)}.$$

It follows that the second sum is upper bounded by $C\varepsilon^u$.
Since $\left\|\partial_j h * \phi_\varepsilon^i\right\|_\infty \leq C \|h\|_\infty$, it follows that

$$d_{k-1}(\mu * \phi_\varepsilon^i, \mu_{\delta_{n,\varepsilon}} * \phi_\varepsilon^i) \leq Cd_k(\mu, \mu_{\delta_{n,\varepsilon}}) \leq \frac{C}{\lambda_{q+1,m,p}(\delta_{n,\varepsilon})^\eta} = \frac{Cn^2}{2^{2mn\eta}}\varepsilon^{\eta(1-u)}.$$

Since $2m\eta > q + k + d/p_*$, the first sum is convergent also, and is upper bounded by $C\varepsilon^{\eta(1-u)}$. We conclude that

$$\left\|\partial_i(g * \phi_\varepsilon^i)\right\|_p \le C\varepsilon^{\eta(1-u)} + C\varepsilon^u.$$

In order to optimize we take $u = \eta/(1+\eta)$. $\qquad\square$

3.3 Random variables and integration by parts

In this section we work in the framework of random variables. For a random variable F we denote by μ_F the law of F, and if μ_F is absolutely continuous we denote by p_F its density. We will use Theorem 3.2.4 for μ_F so we will look for a family of random variables F_δ, $\delta > 0$, such that the associated family of laws $\mu_{F_\delta}, \delta > 0$ satisfies the hypothesis of that theorem. Sometimes it is easy to construct such a family with explicit densities p_{F_δ}, and then (3.18) can be checked directly. But sometimes p_{F_δ} is not known, and then the integration by parts machinery developed in Chapter 1 is quite useful in order to prove (3.18). So, we recall the abstract definition of integration by parts formulas and the useful properties from Chapter 1. We have already applied these results and got estimates that we used in essential manner in Chapter 2. Here we use a variant of these results (specifically, of Theorem 2.3.1) which employs other norms that are more interesting to be handled in this special context. Let us be more precise.

We consider a random vector $F = (F_1, \dots, F_d)$ and a random variable G. Given a multiindex $\alpha = (\alpha_1, \dots, \alpha_k) \in \{1, \dots, d\}^k$ and for $p \ge 1$ we say that $\mathrm{IBP}_{\alpha,p}(F, G)$ holds if one can find a random variable $H_\alpha(F; G) \in L^p$ such that for every $f \in C^\infty(\mathbb{R}^d)$ we have

$$\mathbb{E}\big(\partial_\alpha f(F)G\big) = \mathbb{E}\big(f(F)H_\alpha(F; G)\big). \tag{3.20}$$

The weight $H_\alpha(F; G)$ is not uniquely determined: the one which has the lowest possible variance is $\mathbb{E}(H_\alpha(F; G) \mid \sigma(F))$. This quantity is uniquely determined. So we denote

$$\theta_\alpha(F, G) = \mathbb{E}\big(H_\alpha(F; G) \mid \sigma(F)\big). \tag{3.21}$$

For $m \in \mathbb{N}$ and $p \ge 1$, we denote by $\mathcal{R}_{m,p}$ the class of random vectors $F \in \mathbb{R}^d$ such that $\mathrm{IBP}_{\alpha,p}(F, 1)$ holds for every multiindex α with $|\alpha| \le m$. We define the norm

$$|||F|||_{m,p} = \|F\|_p + \sum_{|\alpha| \le m} \|\theta_\alpha(F, 1)\|_p. \tag{3.22}$$

Notice that, by Hölder's inequality, $\|\mathbb{E}(H_\alpha(F, 1) \mid \sigma(F))\|_p \le \|H_\alpha(F, 1)\|_p$. It follows that, for every choice of the weights $H_\alpha(F, 1)$, we have

$$|||F|||_{m,p} \le \|F\|_p + \sum_{|\alpha| \le m} \|H_\alpha(F, 1)\|_p. \tag{3.23}$$

Theorem 3.3.1. *Let $m \in \mathbb{N}$ and $p > d$. Let $F \in \bigcap_{\bar{p} \geq 1} L^{\bar{p}}(\Omega)$. If $F \in \mathcal{R}_{m+1,p}$, then the law of F is absolutely continuous and the density p_F belongs to $C^m(\mathbb{R}^d)$. Moreover, suppose that $F \in \mathcal{R}_{m+1,2(d+1)}$. Then, there exists a universal constant C (depending on d and m only) such that, for every multiindex α with $|\alpha| \leq m$ and every $k \in \mathbb{N}$,*

$$|\partial_\alpha p_F(x)| \leq C \, |||F|||_{1,2(d+1)}^{d^2-1} \, |||F|||_{m+1,2(d+1)} \left(1 + \|F\|_k^{\frac{k}{2p}} \right) \left(1 + |x| \right)^{-\frac{k}{2p}}. \quad (3.24)$$

In particular, for every $q \geq 1$, $k \in \mathbb{N}$, there exists a universal constant C (depending on d, m, k, and p only) such that

$$\|p_F\|_{m,k,q} \leq C \, |||F|||_{1,2(d+1)}^{d^2-1} \, |||F|||_{m+1,2(d+1)} \left(1 + \|F\|_{4p(k+d+1)}^{2(k+d+1)} \right). \quad (3.25)$$

Proof. The statements follow from the results in Chapter 1. In order to see this we have to explain the relation between the notation used there and the one used here: we work with the probability measure $\mu_F(dx) = \mathbb{P}(F \in dx)$ and in Chapter 1 we used the notation $\partial_\alpha^{\mu_F} g(x) = (-1)^{|\alpha|} \mathbb{E}(H_\alpha(F; g(F)) \mid F = x)$.

The fact $F \in \mathcal{R}_{m+1,p}$ is equivalent to $1 \in W_{\mu_F}^{m+1,p}$ so, the fact that $F \sim p_F(x)dx$ with $p_F \in C^m(\mathbb{R}^d)$ is proved in Theorem 1.5.2 (take there $\phi \equiv 1$). Consider now a function $\psi \in C_b^\infty(\mathbb{R}^d)$ such that $1_{B_1} \leq \psi \leq 1_{B_2}$, where B_r denotes the ball of radius r centered at 0. In Remark 1.6.2 we have given the representation formula

$$\partial_\alpha p_F(x) = \sum_{i=1}^d \mathbb{E}\big(\partial_i Q_d(F - x) \theta_{(\alpha,i)}\big(F; \psi(F - x)\big) 1_{B_2}(F - x) \big),$$

where Q_d is the Poisson kernel on $\mathbb{R}^d$ and, for a multiindex $\alpha = (\alpha_1, \ldots, \alpha_k)$, we set $(\alpha, i) = (\alpha_1, \ldots, \alpha_k, i)$ as usual. Using Hölder's inequality with $p > d$ we obtain

$$|\partial_\alpha p_F(x)| \leq \sum_{i=1}^d \big\| \partial_i Q_d(F - x) \big\|_{p_*} \big\| \theta_{(\alpha,i)}\big(F; \psi(F - x)\big) 1_{B_2}(F - x) \big\|_p,$$

where p_* stands for the conjugate of p. We now take $p = d + 1$ so that $p_* = (d+1)/d \leq 2$, and we apply Theorem 1.4.1: by using the norms in (3.22) and the estimate (3.23), we get

$$\big\| \partial_i Q_d(F - x) \big\|_{p_*} \leq C \, |||F|||_{1,2(d+1)}^{d^2-1}.$$

Moreover, by the computational rules in Proposition 1.1.1, we have

$$\theta_i(F, fg(F)) = f(F)\theta_i(F, g(F)) + (g\partial_i f)(F).$$

Since $\psi \in C_b^\infty(\mathbb{R}^d)$, setting $\phi_x(\cdot) = \phi(\cdot - x)$ the above formula yields

$$\big\| \theta_{(\alpha,i)}\big(F; \psi(F - x)\big) 1_{B_2}(F - x) \big\|_p \leq \big\| \theta_{(\alpha,i)}\big(F; \psi(F - x)\big) \big\|_{2p} \mathbb{P}\big(|F - x| \leq 2 \big)^{\frac{1}{2p}}$$

$$\leq C_\psi \, |||F|||_{|\alpha|+1,2p} \, \mathbb{P}\big(|F - x| \leq 2 \big)^{\frac{1}{2p}}.$$

For $|x| \geq 4$

$$\mathbb{P}\big(|F - x| \leq 2\big) \leq \mathbb{P}\bigg(|F| \geq \frac{1}{2}|x|\bigg) \leq \frac{2^k}{|x|^k}\mathbb{E}(|F|^k)$$

and the proof is complete. □

We are now ready to state Theorem 3.2.4 in terms of random variables.

Theorem 3.3.2. *Let $k, q \in \mathbb{N}$, $m \in \mathbb{N}_*$, $p > 1$, and let p_* denote the conjugate of p. Let F, F_δ, $\delta > 0$, be random variables and let μ_F, μ_{F_δ}, $\delta > 0$, denote the associated laws. Suppose that $F \in \bigcap_{\bar{p} \geq 1} L^{\bar{p}}(\Omega)$.*

(i) *Suppose that $F_\delta \in \mathcal{R}_{2m+q+1,2(d+1)}$, $\delta > 0$, and that there exist $C > 0$ and $\theta \geq 0$ such that*

$$|||F_\delta|||_{n,2(d+1)} \leq C\delta^{-\theta n} \quad \text{for every} \quad n \leq 2m + q + 1, \tag{3.26}$$

$$d_k\big(\mu_F, \mu_{F_\delta}\big) \leq C\delta^{\theta\eta(d^2+q+1+2m)} \quad \text{for some} \quad \eta > \tfrac{q+k+d/p_*}{2m}. \tag{3.27}$$

Then, $\mu_F(dx) = p_F(x)dx$ with $p_F \in W^{q,p}$.

(ii) *Let μ_F, μ_{F_δ}, $\delta > 0$, be such that*

$$\mathbb{E}(|F|) + \sup_\delta \mathbb{E}(|F_\delta|) < \infty. \tag{3.28}$$

Suppose that $F_\delta \in \mathcal{R}_{2m+q+2,2(d+1)}$, $\delta > 0$, and that there exist $C > 0$ and $\theta \geq 0$ such that

$$|||F_\delta|||_{n,2(d+1)} \leq C\delta^{-\theta n} \quad \text{for every} \quad n \leq 2m + q + 2,$$

$$d_k\big(\mu_F, \mu_{F_\delta}\big) \leq C\delta^{\theta\eta(d^2+q+2+2m)} \quad \text{for some} \quad \eta > \tfrac{q+k+d/p_*}{2m}.$$

Then, for every multiindex α with $|\alpha| = q$ and every $s < s_\eta(q, k, m, p)$, we have $\partial_\alpha p_F \in \mathcal{B}^{s,p}$, where $\mathcal{B}^{s,p}$ is the Besov space of index s and $s_\eta(q, k, m, p)$ is given in (3.19).

Proof. Let $n \in \mathbb{N}$, $l \in \mathbb{N}$, and $p > 1$ be fixed. By using (3.25) and (3.26), we obtain $\|p_F\|_{n,l,p} \leq C\delta^{-\theta(d^2+n)}$. So, as a consequence of (3.27) we obtain

$$\|p_F\|_{2m+q,2m,p}^\eta \, d_k\big(\mu_F, \mu_{F_\delta}\big) \leq C.$$

Now the statements follow by applying Theorem 3.2.3 and Theorem 3.2.4. □

Remark 3.3.3. In the previous theorem we assumed that all the random variables F_δ, $\delta > 0$, are defined on the same probability space $(\Omega, \mathcal{F}, \mathbb{P})$. But this is just for simplicity of notations. In fact the statement concerns just the law of $(F_\delta, H_\alpha(F_\delta, 1), |\alpha| \leq m)$, so we may assume that each F_δ is defined on a different probability space $(\Omega_\delta, \mathcal{F}_\delta, \mathbb{P}_\delta)$. This remark is used in [6]: there, we have a unique random variable F on a measurable space $(\Omega, \mathcal{F})$ and we look at the law of F under different probability measures $\mathbb{P}_\delta$, $\delta > 0$, on $(\Omega, \mathcal{F})$.

3.4   Examples

3.4.1   Path dependent SDE's

We denote by $C(\mathbb{R}_+; \mathbb{R}^d)$ the space of continuous functions from $\mathbb{R}_+$ to $\mathbb{R}^d$, and we consider some measurable functions

$$\sigma_j, b \colon C(\mathbb{R}_+; \mathbb{R}^d) \longrightarrow C(\mathbb{R}_+; \mathbb{R}^d), \quad j = 1, \ldots, n$$

which are adapted to the standard filtration given by the coordinate functions. We set

$$\sigma_j(t, \varphi) = \sigma_j(\varphi)_t \quad \text{and} \quad b(t, \varphi) = b(\varphi)_t, \quad j = 1, \ldots, n, \ t \geq 0, \ \varphi \in C(\mathbb{R}_+; \mathbb{R}^d).$$

Then we look at the process solution to the equation

$$dX_t = \sum_{j=1}^{n} \sigma_j(t, X) dW_t^j + b(t, X) dt, \tag{3.29}$$

where $W = (W^1, \ldots, W^n)$ is a standard Brownian motion. If σ_j and b satisfy some Lipschitz continuity property with respect to the supremum norm on $C(\mathbb{R}_+; \mathbb{R}^d)$, then this equation has a unique solution. But we do not want to make such an hypothesis here, so we just consider an adapted process X_t, $t \geq 0$, that verifies (3.29). We assume that σ_j and b are bounded. For $s < t$ we denote

$$\Delta_{s,t}(w) = \sup_{s \leq u \leq t} |w_u - w_s|, \quad w \in C(\mathbb{R}_+; \mathbb{R}^d),$$

and we assume that there exists $h, C > 0$ such that

$$\left| \sigma_j(t, w) - \sigma_j(s, w) \right| \leq C \left| \Delta_{s,t}(w) \right|^h, \quad \forall j = 1, \ldots, n. \tag{3.30}$$

Moreover, we assume that there exists some $\lambda > 0$ such that

$$\sigma \sigma^*(t, w) \geq \lambda \quad \forall t \geq 0, w \in C(\mathbb{R}_+; \mathbb{R}^d). \tag{3.31}$$

Theorem 3.4.1. *Suppose that (3.30) and (3.31) hold. Let $p > 1$ be such that $p - 1 < hp/2d$. Then, for every $T > 0$, the law of X_T is absolutely continuous with respect to the Lebesgue measure and the density p_T belongs to $\mathcal{B}^{s,p}$ for some $s > 0$ depending on p and h.*

Remark 3.4.2. In the particular case of standard SDE's we have $\sigma_j(t, w) = \sigma_j(w_t)$ and a sufficient condition for (3.30) to hold is $|\sigma_j(x) - \sigma_j(y)| \leq C |x - y|^h$.

Remark 3.4.3. The proof we give below works for general Itô processes; however, we restrict ourself to SDE's because this is the most familiar example.

Proof of Theorem 3.4.1. During this proof C is a constant depending on $\|b\|_\infty$, $\|\sigma_j\|_\infty$, and on the constants in (3.30) and (3.31), and which may change from a line to another. We take $\delta > 0$ and define

$$X_T^\delta = X_{T-\delta} + \sum_{j=1}^n \sigma_j(T-\delta, X)\left(W_T^j - W_{T-\delta}^j\right).$$

Using (3.30) it follows that, for $T - \delta \le t \le T$, we have (for small $\delta > 0$)

$$\mathbb{E}\left(\left|X_T - X_T^\delta\right|^2\right) \le C\delta^2 + \int_{T-\delta}^T \mathbb{E}\left(\left|\sigma_j(t,X) - \sigma_j(T-\delta, X)\right|^2\right)dt$$

$$\le C\delta^2 + C\delta\mathbb{E}\left(\Delta_{T-\delta,T}(X)^{2h}\right) \le C\delta^{1+h},$$

so

$$d_1\left(\mu_{X_T}, \mu_{X_T^\delta}\right) \le \mathbb{E}\left(\left|X_T - X_T^\delta\right|\right) \le C\delta^{\frac{1}{2}(1+h)},$$

where μ_{X_T} and $\mu_{X_T^\delta}$ denote the law of X_T and of X_T^δ, respectively.

Moreover, conditionally to $\sigma(W_t, t \le T - \delta)$, the law of X_T^δ is Gaussian with covariance matrix larger than $\lambda > 0$. It follows that the law of X_T^δ is absolutely continuous. And, for every $m \in \mathbb{N}$ and every multiindex α, the density p_δ verifies

$$\|p_\delta\|_{2m+1,2m,p} \le C\delta^{-\frac{1}{2}(2m+1)}.$$

We will use Theorem 3.2.4 with $q = 0$ and $k = 1$. Take $\varepsilon > 0$ and

$$\eta = \frac{1 + d/p_* + \varepsilon}{2m} > \frac{1 + d/p_*}{2m}.$$

Then we obtain

$$\|p_\delta\|_{2m+q+1,2m,p}^\eta \, d_1\left(\mu_{X_T}, \mu_{X_T^\delta}\right) \le C\delta^{\frac{1}{2}(1+h)-\frac{1}{2}(2m+1)\eta},$$

where

$$\theta = \frac{1}{2}(1+h) - \frac{1}{2}(2m+1)\frac{1 + d/p_* + \varepsilon}{2m}.$$

We may take m as large as we want and ε as small as we want so we obtain $\theta \ge h/2 - d/p_*$ and this quantity is strictly larger than zero if $h/2 > d(p-1)/p$. And this is true by our assumptions on p. $\qquad\square$

3.4.2 Diffusion processes

In this subsection we restrict ourselves to standard diffusion processes, thus $\sigma_j, b : \mathbb{R}^d \to \mathbb{R}^d$, $j = 1, \ldots, n$, and the equation is

$$dX_t = \sum_{j=1}^n \sigma_j(X_t)dW_t^j + b(X_t)dt. \tag{3.32}$$

We consider $h > 0$ and $q \in \mathbb{N}$ and we denote by $C_b^{q,h}(\mathbb{R}^d)$ the space of those functions which are q times differentiable, bounded, and with bounded derivatives, and the derivatives of order q are Hölder continuous of order h.

Theorem 3.4.4. *Suppose that $\sigma_j \in C_b^{q,h}(\mathbb{R}^d)$, $j = 1, \ldots, n$, and $b \in C_b^{q-1,h}(\mathbb{R}^d)$ for some $q \in \mathbb{N}$ (in the case $q = 0$, σ_j is just Hölder continuous of index h, and b is just measurable and bounded). Suppose also that*

$$\sigma\sigma^*(x) \geq \lambda > 0 \quad \forall x \in \mathbb{R}^d. \tag{3.33}$$

Then for every $T > 0$ the law of X_T is absolutely continuous with respect to the Lebesgue measure, and the density p_T belongs to $W^{q,p}$ with $1 < p < d/(d - h)$. Moreover, there exists s_* (depending on q, p and which may be computed explicitly) such that, for every multiindex α of length q, we have $\partial_\alpha p_T \in \mathcal{B}^{s,p}$ for $s < s_*$.

Proof. We treat only the case $q = 1$ (the general case is analogous). During this proof C is a constant which depends on the upper bounds of σ_j, b, and their derivatives, on the Hölder constant, and on the constants in (3.33). We take $\delta > 0$ and define

$$X_T^\delta = X_{T-\delta} + \sum_{j=1}^n \sigma_j(X_{T-\delta})(W_T^j - W_{T-\delta}^j)$$

$$+ \sum_{j,k=1}^n \partial_{\sigma_k}\sigma_j(X_{T-\delta}) \int_{T-\delta}^T (W_t^j - W_{T-\delta}^j)dW_t^k + b(X_{T-\delta})\delta,$$

with $\partial_{\sigma_k}\sigma_j = \sum_{i=1}^d \sigma_k^i \partial_i \sigma_j$. Then,

$$X_T - X_T^\delta = \sum_{j,k=1}^n \int_{T-\delta}^T \int_{T-\delta}^t (\partial_{\sigma_k}\sigma_j(X_s) - \partial_{\sigma_k}\sigma_j(X_{T-\delta}))dW_s^k dW_t^j$$

$$+ \sum_{j=1}^n \int_{T-\delta}^T \int_{T-\delta}^t (\partial_b \sigma_j(X_s) - \partial_b \sigma_j(X_{T-\delta}))dsdW_t^j$$

$$+ \int_{T-\delta}^T (b(X_t) - b(X_{T-\delta}))dt.$$

It follows that

$$\mathbb{E}(|X_T - X_T^\delta|^2) \leq C \int_{T-\delta}^T \int_{T-\delta}^t \mathbb{E}(|X_s - X_{T-\delta}|^{2h})dsdt$$

$$+ C\delta \int_{T-\delta}^T \mathbb{E}(|X_t - X_{T-\delta}|^{2h})dt$$

$$\leq C\delta^2 \mathbb{E}\Big(\sup_{T-\delta \leq t \leq T} |X_t - X_{T-\delta}|^{2h}\Big) \leq C\delta^{2+h}$$

and, consequently,

$$d_1\left(X_T, X_T^\delta\right) \le C\delta^{1+h/2}.$$

We have to control the density of the law of X_T^δ. But now we no longer have an explicit form for this density (as in the proof of the previous example) so, we have to use integration by parts formulas built using Malliavin calculus for $W_t - W_{T-\delta}$, $t \in (T - \delta, T)$, conditionally to $\sigma(W_t, t \le T - \delta)$. This is a standard easy exercise, and we obtain

$$\mathbb{E}(\partial_\alpha f(X_T^\delta)) = \mathbb{E}(f(X_T^\delta)H_\alpha(X_T, 1)^\delta)),$$

where $H_\alpha(X_T^\delta, 1)$ is a suitable random variable which verifies

$$\left\|H_\alpha(X_T^\delta, 1)\right\|_p \le C\delta^{-\frac{1}{2}|\alpha|}.$$

Here, δ appears because we use Malliavin calculus on an interval of length δ. The above estimate holds for each $p \ge 1$ and for each multiindex α (notice that, since we work conditionally to $\sigma(W_t, t \le T - \delta)$, $\sigma_j(X_{T-\delta})$ and $b(X_{T-\delta})$ are not involved in the calculus —they act as constants). We conclude that X_T^δ verifies (3.28) and (3.26). We fix now $p > 1$ and $m \in \mathbb{N}$. Then the hypothesis (3.27) (with $q = 1$ and $k = 1$) reads

$$\delta^{1+\frac{h}{2}} \le C\delta^{\frac{\eta}{2}(d^2+1+2m)}, \quad \text{with} \quad \eta > \frac{2 + \frac{d}{p_*}}{2m}.$$

This is possible if

$$1 + \frac{h}{2} > \left(1 + \frac{d}{2p_*}\right)\left(1 + \frac{d^2+1}{2m}\right).$$

Since we can choose m as large as we want, the above inequality can be achieved if $p > 1$ is sufficiently small in order to have $h > d/p_*$. Now, we use Theorem 3.3.2 and we conclude. $\qquad\square$

Remark 3.4.5. In the proof of Theorem 3.4.4 we do not have an explicit expression of the density of the law of X_T^δ so, this is a situation where the statement of our criterion in terms of integration by parts becomes useful.

Remark 3.4.6. The results in the previous theorems may be proved for diffusions with boundary conditions and under a local regularity hypothesis. Hörmander non-degeneracy conditions may also be considered, see [6]. Since this developments are rather technical, we leave them out here.

3.4.3 Stochastic heat equation

In this section we investigate the regularity of the law of the solution to the stochastic heat equation introduced by Walsh in [48]. Formally this equation is

$$\partial_t u(t, x) = \partial_x^2 u(t, x) + \sigma(u(t, x))W(t, x) + b(u(t, x)), \qquad (3.34)$$

where W denotes a white noise on $\mathbb{R}_+ \times [0,1]$. We consider Neumann bound-ary conditions, that is $\partial_x u(t, 0) = \partial_x u(t, 1) = 0$, and the initial condition is $u(0, x) = u_0(x)$. The rigorous formulation to this equation is given by the mild form constructed as follows. Let $G_t(x, y)$ be the fundamental solution to the deterministic heat equation $\partial_t v(t, x) = \partial_x^2 v(t, x)$ with Neumann boundary conditions. Then u satisfies

$$u(t, x) = \int_0^1 G_t(x, y) u_0(y) dy + \int_0^t \int_0^1 G_{t-s}(x, y) \sigma(u(s, y)) dW(s, y) \qquad (3.35)$$

$$+ \int_0^t \int_0^1 G_{t-s}(x, y) b(u(s, y)) dy ds,$$

where $dW(s, y)$ is the Itô integral introduced by Walsh. The function $G_t(x, y)$ is explicitly known (see [11, 48]), but here we will use just the few properties listed below (see the appendix in [11] for the proof). More precisely, for $0 < \varepsilon < t$, we have

$$\int_{t-\varepsilon}^t \int_0^1 G_{t-s}^2(x, y) dy ds \leq C\varepsilon^{1/2}. \qquad (3.36)$$

Moreover, for $0 < x_1 < \cdots < x_d < 1$ there exists a constant C depending on $\min_{i=1,\dots,d}(x_i - x_{i-1})$ such that

$$C\varepsilon^{1/2} \geq \inf_{|\xi|=1} \int_{t-\varepsilon}^t \int_0^1 \left(\sum_{i=1}^d \xi_i G_{t-s}(x_i, y) \right)^2 dy ds \geq C^{-1}\varepsilon^{1/2}. \qquad (3.37)$$

This is an easy consequence of the inequalities (A2) and (A3) from [11].

In [42], sufficient conditions for the absolute continuity of the law of $u(t, x)$ for $(t, x) \in (0, \infty) \times [0, 1]$ are given; and, in [11], under appropriate hypotheses, a C^∞ density for the law of the vector $(u(t, x_1), \dots, u(t, x_d))$ with $(t, x_i) \in (0, \infty) \times \{\sigma \neq 0\}$, $i = 1, \dots, d$, is obtained. The aim of this subsection is to obtain the same type of results, but under much weaker regularity hypothesis on the coefficients. We can discuss, first, the absolute continuity of the law and, further, under more regularity hypotheses on the coefficients, the regularity of the density. Here, in order to avoid technicalities, we restrict ourself to the absolute continuity property. We also assume global ellipticity, i.e.,

$$\sigma(x) \geq c_\sigma > 0 \quad \text{for every } x \in [0, 1]. \qquad (3.38)$$

A local ellipticity condition may also be used but, again, this gives more technical complications that we want to avoid. This is somehow a benchmark for the efficiency of the method developed in the previous sections.

We assume the following regularity hypothesis: σ, b are measurable and bounded functions and there exists $h, C > 0$ such that

$$|\sigma(x) - \sigma(y)| \leq C|x - y|^h, \quad \text{for every } x, y \in [0, 1]. \qquad (3.39)$$

This hypothesis is not sufficient in order to ensure existence and uniqueness for the solution to (3.35) (one also needs σ and b to be globally Lipschitz continuous). So, in the following we will just consider a random field $u(t,x), (t,x) \in (0,\infty) \times [0,1]$ which is adapted to the filtration generated by W (see Walsh [48] for precise definitions) and which solves (3.35).

Proposition 3.4.7. *Suppose that* (3.38) *and* (3.39) *hold. Then, for every* $0 < x_1 < \cdots < x_d < 1$ *and* $T > 0$, *the law of the random vector* $U = (u(T,x_1),\ldots,u(T,x_d))$ *is absolutely continuous with respect to the Lebesgue measure. Moreover, if* $p > 1$ *is such that* $p - 1 < hp/d$, *then the density* p_U *belongs to the Besov space* $\mathcal{B}^{s,p}$ *for some* $s > 0$ *depending on* h *and* p.

Proof. Given $0 < \varepsilon < T$, we decompose

$$u(T,x) = u_\varepsilon(T,x) + I_\varepsilon(T,x) + J_\varepsilon(T,x), \tag{3.40}$$

with

$$u_\varepsilon(T,x) = \int_0^1 G_t(x,y)u_0(y)dy + \int_0^T \int_0^1 G_{T-s}(x,y)\sigma\big(u(s \wedge (T-\varepsilon),y)\big)dW(s,y)$$
$$+ \int_0^{T-\varepsilon} \int_0^1 G_{T-s}(x,y)b(u(s,y))dyds,$$

$$I_\varepsilon(T,x) = \int_{T-\varepsilon}^T \int_0^1 G_{T-s}(x,y)\big(\sigma(u(s,y)) - \sigma(u(s \wedge (T-\varepsilon),y))\big)dW(s,y),$$

$$J_\varepsilon(T,x) = \int_{T-\varepsilon}^T \int_0^1 G_{T-s}(x,y)b(u(s,y))dyds.$$

Step 1. Using the isometry property (3.39) and (3.36),

$$\mathbb{E}|I_\varepsilon(T,x)|^2 = \int_{T-\varepsilon}^T \int_0^1 G_{T-s}^2(x,y)\mathbb{E}\big(\sigma(u(s,y) - \sigma(u(s \wedge (T-\varepsilon),y)))^2\big)dyds$$
$$\leq C \int_{T-\varepsilon}^T \int_0^1 G_{T-s}^2(x,y)\mathbb{E}\Big(\sup_{T-\delta \leq s \leq T}|u(s,y) - u(T-\delta,y)|^{2h}\Big)dyds$$
$$\leq C\varepsilon^{\frac{1}{2}(1+h)},$$

where the last inequality is a consequence of the Hölder property of $u(s,y)$ with respect to s. One also has

$$\mathbb{E}|J_\varepsilon(T,x)|^2 \leq \|b\|_\infty^2 \left(\int_{T-\varepsilon}^T \int_0^1 G_{T-s}(x,y)dyds\right)^2 \leq C\varepsilon^2$$

so, finally,

$$\mathbb{E}|u(T,x) - u_\varepsilon(T,x)|^2 \leq C\varepsilon^{\frac{1}{2}(1+h)}.$$

Let μ be the law of $U = (u(T, x_1), \ldots, u(T, x_d))$ and μ_ε be the law of $U_\varepsilon = (u_\varepsilon(T, x_1), \ldots, u_\varepsilon(T, x_d))$. Using the above estimate we obtain

$$d_1(\mu, \mu_\varepsilon) \leq C \varepsilon^{\frac{1}{4}(1+h)}. \tag{3.41}$$

Step 2. Conditionally to $\mathcal{F}_{T-\varepsilon}$, the random vector $U_\varepsilon = (u_\varepsilon(T, x_1), \ldots, u_\varepsilon(T, x_d))$ is Gaussian with covariance matrix

$$\Sigma^{i,j}(U_\varepsilon) = \int_{T-\varepsilon}^T \int_0^1 G_{T-s}(x_i, y) G_{T-s}(x_j, y) \sigma^2\big(u(s \wedge (T-\varepsilon), y)\big) dy ds,$$

for $i, j = 1, \ldots, d$. By (3.37),

$$C\sqrt{\varepsilon} \geq \Sigma(U_\varepsilon) \geq \frac{1}{C}\sqrt{\varepsilon}$$

where C is a constant depending on the upper bounds of σ, and on c_σ.

We use now the criterion given in Theorem 3.2.4 with $k = 1$ and $q = 0$. Let p_{U_ε} be the density of the law of U_ε. Conditionally to $\mathcal{F}_{T-\varepsilon}$, this is a Gaussian density and direct estimates give

$$\|p_{U_\varepsilon}\|_{2m+1,2m,p} \leq C \varepsilon^{-(2m+1)/4} \quad \forall m \in \mathbb{N}.$$

We now take $\varepsilon > 0$ and $\eta = (1 + d/p_* + \varepsilon)/2m > (1 + d/p_*)/2m$, and we write

$$\|p_{U_\varepsilon}\|_{2m+1,2m,p}^\eta \, d_1(\mu, \mu_\varepsilon) \leq C \varepsilon^{(1+h)/4 - \eta(2m+1)/4}.$$

Finally, for sufficiently large $m \in \mathbb{N}$ and sufficiently small $\varepsilon > 0$, we have

$$\frac{1}{4}(1+h) - \frac{1}{4}\eta(2m+1) \geq \frac{1}{4}(1+h) - \frac{2m+1}{2m} \times \frac{1 + d/p_* + \varepsilon}{4} \geq 0. \qquad \square$$

3.5 Appendix A: Hermite expansions and density estimates

The aim of this Appendix is to give the proof of Proposition 3.2.1. Actually, we are going to prove more. Recall that, for $\mu \in \mathcal{M}$ and $\mu_n(x) = f_n(x)dx$, $n \in \mathbb{N}$, and for $p > 0$ (with p_* the conjugate of p) we consider

$$\pi_{q,k,m,p}(\mu, (\mu_n)_n) = \sum_{n=0}^\infty 2^{n(q+k+d/p_*)} d_k(\mu, \mu_n) + \sum_{n=0}^\infty \frac{1}{2^{2nm}} \|f_n\|_{2m+q,2m,p}.$$

Our goal is to prove the following result.

Proposition 3.5.1. *Let $q, k \in \mathbb{N}$, $m \in \mathbb{N}_*$, and $p > 1$. There exists a universal constant C (depending on q, k, m, d, and p) such that, for every $f, f_n \in C^{2m+q}(\mathbb{R}^d)$, $n \in \mathbb{N}$, one has*

$$\|f\|_{q,p} \leq C \pi_{q,k,m,p}(\mu, (\mu_n)_n), \tag{3.42}$$

where $\mu(x) = f(x)dx$ and $\mu_n(x) = f_n(x)dx$.

The proof of Proposition 3.5.1 will follow from the next results and properties of Hermite polynomials, so we postpone it to the end of this appendix.

We begin with a review of some basic properties of Hermite polynomials and functions. The Hermite polynomials on $\mathbb{R}$ are defined by

$$H_n(t) = (-1)^n e^{t^2} \frac{d^n}{dt^n} e^{-t^2}, \quad n = 0, 1, \ldots$$

They are orthogonal with respect to $e^{-t^2} dt$. We denote the L^2 normalized Hermite functions by

$$h_n(t) = \left(2^n n! \sqrt{\pi}\right)^{-1/2} H_n(t) e^{-t^2/2}$$

and we have

$$\int_{\mathbb{R}} h_n(t) h_m(t) dt = (2^n n! \sqrt{\pi})^{-1} \int_{\mathbb{R}} H_n(t) H_m(t) e^{-t^2} dt = \delta_{n,m}.$$

The Hermite functions form an orthonormal basis in $L^2(\mathbb{R})$. For a multiindex $\alpha = (\alpha_1, \ldots, \alpha_d) \in \mathbb{N}^d$, we define the d-dimensional Hermite function as

$$\mathcal{H}_\alpha(x) := \prod_{i=1}^{d} h_{\alpha_i}(x_i), \quad x = (x_1, \ldots, x_d).$$

The d-dimensional Hermite functions form an orthonormal basis in $L^2(\mathbb{R}^d)$. This corresponds to the chaos decomposition in dimension d (but the notation we gave above is slightly different from the one customary in probability theory; see [33, 41, 44], where Hermite polynomials are used. We can come back by a renormalization). The Hermite functions are the eigenvectors of the Hermite operator $D = -\Delta + |x|^2$, where Δ denotes the Laplace operator. We have

$$D\mathcal{H}_\alpha = (2|\alpha| + d)\mathcal{H}_\alpha, \quad \text{with} \quad |\alpha| = \alpha_1 + \cdots + \alpha_d. \qquad (3.43)$$

We denote $W_n = \mathrm{Span}\{\mathcal{H}_\alpha : |\alpha| = n\}$ and we have $L^2(\mathbb{R}^d) = \bigoplus_{n=0}^{\infty} W_n$.

For a function $\Phi \colon \mathbb{R}^d \times \mathbb{R}^d \to \mathbb{R}$ and a function $f \colon \mathbb{R}^d \to \mathbb{R}$, we use the notation

$$\Phi \diamond f(x) = \int_{\mathbb{R}^d} \Phi(x, y) f(y) dy.$$

We denote by J_n the orthogonal projection on W_n, and then we have

$$J_n v(x) = \bar{\mathcal{H}}_n \diamond v(x) \quad \text{with} \quad \bar{\mathcal{H}}_n(x, y) := \sum_{|\alpha|=n} \mathcal{H}_\alpha(x) \mathcal{H}_\alpha(y). \qquad (3.44)$$

Moreover, we consider a function $a \colon \mathbb{R}_+ \to \mathbb{R}$ whose support is included in $[\frac{1}{4}, 4]$ and we define

$$\bar{\mathcal{H}}_n^a(x, y) = \sum_{j=0}^{\infty} a\left(\frac{j}{4^n}\right) \bar{\mathcal{H}}_j(x, y) = \sum_{j=4^{n-1}+1}^{4^{n+1}-1} a\left(\frac{j}{4^n}\right) \bar{\mathcal{H}}_j(x, y), \quad x, y \in \mathbb{R}^d,$$

the last equality being a consequence of the assumption on the support of a.

The following estimate is a crucial point in our approach. It has been proved in [23, 24, 43] (we refer to [43, Corollary 2.3, inequality (2.17)]).

Theorem 3.5.2. *Let $a: \mathbb{R}_+ \to \mathbb{R}_+$ be a nonnegative C^∞ function with the support included in $[1/4, 4]$. We denote $\|a\|_l = \sum_{i=0}^l \sup_{t \geq 0} |a^{(i)}(t)|$. For every multiindex α and every $k \in \mathbb{N}$ there exists a constant C_k (depending on k, α, d) such that, for every $n \in \mathbb{N}$ and every $x, y \in \mathbb{R}^d$,*

$$\left| \frac{\partial^{|\alpha|}}{\partial x^\alpha} \bar{\mathcal{H}}_n^a(x, y) \right| \leq C_k \|a\|_k \frac{2^{n(|\alpha|+d)}}{(1 + 2^n |x - y|)^k}. \tag{3.45}$$

Following the ideas in [43], we consider a function $a: \mathbb{R}_+ \to \mathbb{R}_+$ of class C_b^∞ with the support included in $[1/4, 4]$ and such that $a(t) + a(4t) = 1$ for $t \in [1/4, 1]$. We construct a in the following way: we take a function $a: [0, 1] \to \mathbb{R}_+$ with $a(t) = 0$ for $t \leq 1/4$ and $a(1) = 1$. We choose a such that $a^{(l)}(1/4^+) = a^{(l)}(1^-) = 0$ for every $l \in \mathbb{N}$. Then, we define $a(t) = 1 - a(t/4)$ for $t \in [1, 4]$ and $a(t) = 0$ for $t \geq 4$. This is the function we will use in the following. Notice that a has the property:

$$\sum_{n=0}^\infty a\left(\frac{t}{4^n}\right) = 1 \quad \forall t \geq 1. \tag{3.46}$$

In order to check this we fix n_t such that $4^{n_t - 1} \leq t < 4^{n_t}$ and we notice that $a(t/4^n) = 0$ if $n \notin \{n_t - 1, n_t\}$. So, $\sum_{n=0}^\infty a(t/4^n) = a(4s) + a(s) = 1$, with $s = t/4^{n_t} \in [\frac{1}{4}, 1]$. In the following we fix a function a and the constants in our estimates will depend on $\|a\|_l$ for some fixed l. Using this function we obtain the following representation formula:

Proposition 3.5.3. *For every $f \in L^2(\mathbb{R}^d)$ one has that*

$$f = \sum_{n=0}^\infty \bar{\mathcal{H}}_n^a \diamond f,$$

the series being convergent in $L^2(\mathbb{R}^d)$.

Proof. We fix N and denote

$$S_N^a = \sum_{n=1}^N \bar{\mathcal{H}}_n^a \diamond f, \quad S_N = \sum_{j=1}^{4^N} \bar{\mathcal{H}}_j \diamond f, \quad and \quad R_N^a = \sum_{j=4^N+1}^{4^{N+1}} (\bar{\mathcal{H}}_j \diamond f) a\left(\frac{j}{4^{N+1}}\right).$$

Let $j \leq 4^{N+1}$. For $n \geq N + 2$ we have $a(j/4^n) = 0$. So, using (3.46), we obtain

$$\sum_{n=1}^N a\left(\frac{j}{4^n}\right) = \sum_{n=1}^\infty a\left(\frac{j}{4^n}\right) - a\left(\frac{j}{4^{N+1}}\right) = 1 - a\left(\frac{j}{4^{N+1}}\right).$$

Moreover, for $j \le 4^N$, we have $a(j/4^{N+1}) = 0$. It follows that

$$
S_N^a = \sum_{n=1}^{N} \sum_{j=0}^{\infty} a\left(\frac{j}{4^n}\right) \bar{\mathcal{H}}_j \diamond f = \sum_{n=1}^{N} \sum_{j=0}^{4^{N+1}} a\left(\frac{j}{4^n}\right) \bar{\mathcal{H}}_j \diamond f = \sum_{j=0}^{4^{N+1}} (\bar{\mathcal{H}}_j \diamond f) \sum_{n=1}^{N} a\left(\frac{j}{4^n}\right)
$$

$$
= \sum_{j=0}^{4^{N+1}} \bar{\mathcal{H}}_j \diamond f - \sum_{j=4^N+1}^{4^{N+1}} (\bar{\mathcal{H}}_j \diamond f) \, a\left(\frac{j}{4^{N+1}}\right) = S_{N+1} - R_N^a.
$$

One has $S_N \to f$ in L^2 and $\|R_N^a\|_2 \le \|a\|_\infty \sum_{j=4^N+1}^{4^{N+1}} \|\bar{\mathcal{H}}_j \diamond f\|_2 \to 0$, so the proof is complete.   □

In the following we consider the function $\rho_n(z) = (1 + 2^n |z|)^{-(d+1)}$, and we will use the following properties (recall that $*$ denotes convolution):

$$
\|\rho_n * f\|_p \le \|\rho_n\|_1 \|f\|_p \le \frac{C}{2^{nd}} \|f\|_p \quad \text{and} \quad \|\rho_n\|_p \le \frac{C}{2^{nd/p}}. \tag{3.47}
$$

Proposition 3.5.4. *Let $p > 1$ and let p_* be the conjugate of p. Let α be a multiindex.*

(i) *There exists a universal constant C (depending on α, d, and p) such that*

    a) $\left\|\partial_\alpha \bar{\mathcal{H}}_n^a \diamond f\right\|_p \le C \|a\|_{d+1} \times 2^{n|\alpha|} \|f\|_p,$

    b) $\left\|\partial_\alpha \bar{\mathcal{H}}_n^a \diamond f\right\|_\infty \le C \|a\|_{d+1} \times 2^{n(|\alpha|+d/p_*)} \|f\|_{p_*}.$ (3.48)

(ii) *Let $m \in \mathbb{N}_*$. There exists a universal constant C (depending on α, m, d, and p) such that*

$$
\left\|\bar{\mathcal{H}}_n^a \diamond \partial_\alpha f\right\|_p \le \frac{C \|a\|_{d+1}^2}{4^{nm}} \|f\|_{2m+|\alpha|, 2m, p}. \tag{3.49}
$$

(iii) *Let $k \in \mathbb{N}$. There exists a universal constant C (depending on α, k, d, and p) such that*

$$
\left\|\bar{\mathcal{H}}_n^a \diamond \partial_\alpha (f - g)\right\|_p \le C \|a\|_{d+1} \times 2^{n(|\alpha|+k+d/p_*)} d_k(\mu_f, \mu_g). \tag{3.50}
$$

Proof. (i) Using (3.45) with $k = d + 1$, we get

$$
\left|\partial_\alpha \bar{\mathcal{H}}_n^a \diamond f(x)\right| \le C 2^{n(|\alpha|+d)} \|a\|_{d+1} \int \rho_n(x - y) |f(y)| \, dy. \tag{3.51}
$$

Then, using (3.47), we obtain

$$
\left\|\partial_\alpha \bar{\mathcal{H}}_n^a \diamond f\right\|_p \le C 2^{n(|\alpha|+d)} \|a\|_{d+1} \|\rho_n * |f|\|_p \le C 2^{n(|\alpha|+d)} \|a\|_{d+1} \|\rho_n\|_1 \|f\|_p
$$

$$
\le C 2^{n|\alpha|} \|a\|_{d+1} \|f\|_p
$$

and (a) is proved. Again by (3.51) and using Hölder's inequality, we obtain

$$\left|\partial_\alpha \bar{\mathcal{H}}_n^a \diamond f(x)\right| \leq C \left\|a\right\|_{d+1} 2^{n(|\alpha|+d)} \int \rho_n(x-y) \left|f(y)\right| dy$$

$$\leq C \left\|a\right\|_{d+1} 2^{n(|\alpha|+d)} \left\|\rho_n\right\|_p \left\|f\right\|_{p_*}.$$

Using the second inequality in (3.47), (b) is proved as well.

(ii) We define the functions $a_m(t) = a(t)t^{-m}$. Since $a(t) = 0$ for $t \leq 1/4$ and for $t \geq 4$, we have $\left\|a_m\right\|_{d+1} \leq C_{m,d} \left\|a\right\|_{d+1}$. Moreover, $D\bar{\mathcal{H}}_j \diamond v = (2j+d)\bar{\mathcal{H}}_j \diamond v$, so we obtain

$$\bar{\mathcal{H}}_j \diamond v = \frac{1}{2j}(D-d)\bar{\mathcal{H}}_j \diamond v.$$

We denote $L_{m,\alpha} = (D-d)^m \partial_\alpha$. Notice that $L_{m,\alpha} = \sum_{|\beta| \leq 2m} \sum_{|\gamma| \leq 2m+|\alpha|} c_{\beta,\gamma} x^\beta \partial_\gamma$, where $c_{\beta,\gamma}$ are universal constants. It follows that there exists a universal constant C such that

$$\left\|L_{m,\alpha} f\right\|_{(p)} \leq C \left\|f\right\|_{2m+|\alpha|,2m,p}. \tag{3.52}$$

We take now $v \in L^{p_*}$ and write

$$\left\langle v, \bar{\mathcal{H}}_n^a \diamond (\partial_\alpha f) \right\rangle = \left\langle \bar{\mathcal{H}}_n^a \diamond v, \partial_\alpha f \right\rangle = \sum_{j=0}^\infty a\left(\frac{j}{4^n}\right) \left\langle \bar{\mathcal{H}}_j \diamond v, \partial_\alpha f \right\rangle$$

$$= \sum_{j=1}^\infty a\left(\frac{j}{4^n}\right) \frac{1}{(2j)^m} \left\langle (D-d)^m \bar{\mathcal{H}}_j \diamond v, \partial_\alpha f \right\rangle$$

$$= \frac{1}{2^m} \times \frac{1}{4^{nm}} \sum_{j=1}^\infty a_m\left(\frac{j}{4^n}\right) \left\langle \bar{\mathcal{H}}_j \diamond v, L_{m,\alpha} f \right\rangle$$

$$= \frac{1}{2^m} \times \frac{1}{4^{nm}} \left\langle \bar{\mathcal{H}}_n^{a_m} \diamond v, L_{m,\alpha} f \right\rangle.$$

Using the decomposition in Proposition 3.5.3, we write $L_{m,\alpha} f = \sum_{j=0}^\infty \bar{\mathcal{H}}_j^a \diamond L_{m,\alpha} f$. For $j < n-1$ and for $j > n+1$ we have $\left\langle \bar{\mathcal{H}}_n^{a_m} \diamond v, \bar{\mathcal{H}}_j^a \diamond L_{m,\alpha} f \right\rangle = 0$. So, using Hölder's inequality,

$$\left|\left\langle v, \bar{\mathcal{H}}_n^a \diamond (\partial_\alpha f) \right\rangle\right| \leq \frac{1}{4^{nm}} \sum_{j=n-1}^{n+1} \left|\left\langle \bar{\mathcal{H}}_n^{a_m} \diamond v, \bar{\mathcal{H}}_j^a \diamond L_{m,\alpha} f \right\rangle\right|$$

$$\leq \frac{1}{4^{nm}} \sum_{j=n-1}^{n+1} \left\|\bar{\mathcal{H}}_n^{a_m} \diamond v\right\|_{p_*} \left\|\bar{\mathcal{H}}_j^a \diamond L_{m,\alpha} f\right\|_p.$$

Using item (i)(a) with α equal to the empty index, we obtain

$$\left\|\bar{\mathcal{H}}_n^{a_m} \diamond v\right\|_{p_*} \leq C \left\|a_m\right\|_{d+1} \left\|v\right\|_{p_*} \leq C \times C_{m,d} \left\|a\right\|_{d+1} \left\|v\right\|_{p_*}.$$

Moreover, we have

$$\left\|\bar{\mathcal{H}}_j^a \diamond L_{m,\alpha} f\right\|_p \leq C \left\|a\right\|_{d+1} \left\|L_{m,\alpha} f\right\|_p \leq C \left\|a\right\|_{d+1} \left\|f\right\|_{2m+|\alpha|,2m,p},$$

the last inequality being a consequence of (3.52). We obtain

$$\left|\langle v, \bar{\mathcal{H}}_n^a \diamond (\partial_\alpha f)\rangle\right| \leq \frac{C \left\|a\right\|_{d+1}^2}{4^{nm}} \left\|v\right\|_{p_*} \left\|f\right\|_{2m+|\alpha|,2m,p}$$

and, since L^p is reflexive, (3.49) is proved.

(iii) Write

$$\left|\langle v, \bar{\mathcal{H}}_n^a \diamond (\partial_\alpha (f-g))\rangle\right| = \left|\langle \bar{\mathcal{H}}_n^a \diamond v, \partial_\alpha (f-g)\rangle\right| = \left|\langle \partial_\alpha \bar{\mathcal{H}}_n^a \diamond v, f-g\rangle\right|$$

$$= \left|\int \partial_\alpha \bar{\mathcal{H}}_n^a \diamond v d\mu_f - \int \partial_\alpha \bar{\mathcal{H}}_n^a \diamond v d\mu_g\right|.$$

We use (3.48)(b) and get

$$\left|\int \partial_\alpha \bar{\mathcal{H}}_n^a \diamond v d\mu_f - \int \partial_\alpha \bar{\mathcal{H}}_n^a \diamond v d\mu_g\right| \leq \left\|\partial_\alpha \bar{\mathcal{H}}_n^a \diamond v\right\|_{k,\infty} d_k(\mu_f, \mu_g)$$

$$\leq \left\|\bar{\mathcal{H}}_n^a \diamond v\right\|_{k+|\alpha|,\infty} d_k(\mu_f, \mu_g)$$

$$\leq C \left\|a\right\|_{d+1} 2^{n(k+|\alpha|+d/p_*)} \left\|v\right\|_{p_*} d_k(\mu_f, \mu_g),$$

which implies (3.50). $\qquad\square$

We are now ready to prove Proposition 3.5.1.

Proof of Proposition 3.5.1. Let α be a multiindex with $|\alpha| \leq q$. Using Proposition 3.5.3,

$$\partial_\alpha f = \sum_{n=1}^\infty \bar{\mathcal{H}}_n^a \diamond \partial_\alpha f = \sum_{n=1}^\infty \bar{\mathcal{H}}_n^a \diamond \partial_\alpha (f - f_n) + \sum_{n=1}^\infty \bar{\mathcal{H}}_n^a \diamond \partial_\alpha f_n$$

and, using (3.50) and (3.49),

$$\left\|\partial_\alpha f\right\|_p \leq \sum_{n=1}^\infty \left\|\bar{\mathcal{H}}_n^a \diamond \partial_\alpha (f - f_n)\right\|_p + \sum_{n=1}^\infty \left\|\bar{\mathcal{H}}_n^a \diamond \partial_\alpha f_n\right\|_p$$

$$\leq C \sum_{n=1}^\infty 2^{n(|\alpha|+k+d/p_*)} d_k(\mu_f, \mu_{f_n}) + C \sum_{n=1}^\infty \frac{1}{2^{2nm}} \left\|f_n\right\|_{2m+|\alpha|,2m,p}.$$

Hence, (3.42) is proved. $\qquad\square$

3.6 Appendix B: Interpolation spaces

In this Appendix we prove that the space $S_{q,k,m,p}$ is an interpolation space between $W_*^{k,\infty}$ (the dual of $W^{k,\infty}$) and $W^{q,2m,p}$. To begin with, we recall the framework of interpolation spaces.

We are given two Banach spaces $(X, \|\cdot\|_X), (Y, \|\cdot\|_Y)$ with $X \subset Y$ (with continuous embedding). We denote by $\mathcal{L}(X, X)$ the space of linearly bounded operators from X into itself, and we denote by $\|L\|_{X,X}$ the operator norm. A Banach space $(W, \|\cdot\|_W)$ such that $X \subset W \subset Y$ (with continuous embedding) is called an interpolation space for X and Y if $\mathcal{L}(X, X) \cap \mathcal{L}(Y, Y) \subset \mathcal{L}(W, W)$ (with continuous embedding). Let $\gamma \in (0, 1)$. If there exists a constant C such that $\|L\|_{W,W} \leq C \|L\|_{X,X}^{\gamma} \|L\|_{Y,Y}^{1-\gamma}$ for every $L \in \mathcal{L}(X, X) \cap \mathcal{L}(Y, Y)$, then W is an interpolation space of order γ. And if we can take $C = 1$, then W is an exact interpolation space of order γ. There are several methods for constructing interpolation spaces. We focus here on the so-called K-method. For $y \in Y$ and $t > 0$ define $K(y, t) = \inf_{x \in X}(\|y - x\|_Y + t \|x\|_X)$ and

$$\|y\|_{\gamma} = \int_0^{\infty} t^{-\gamma} K(y, t) \frac{dt}{t}, \quad (X, Y)_{\gamma} = \{y \in Y : \|y\|_{\gamma} < \infty\}.$$

Then it is proven that $(X, Y)_{\gamma}$ is an exact interpolation space of order γ. One may also use the following discrete variant of the above norm. Let $\gamma \geq 0$. For $y \in Y$ and for a sequence $x_n \in X$, $n \in \mathbb{N}$, we define

$$\pi_{\gamma}(y, (x_n)_n) = \sum_{n=1}^{\infty} 2^{n\gamma} \|y - x_n\|_Y + \frac{1}{2^n} \|x_n\|_X \qquad (3.53)$$

and

$$\rho_{\gamma}^{X,Y}(y) = \inf \pi_{\gamma}(y, (x_n)_n),$$

with the infimum taken over all the sequences $x_n \in X$, $n \in \mathbb{N}$. Then a standard result in interpolation theory (the proof is elementary) says that there exists a constant $C > 0$ such that

$$\frac{1}{C} \|y\|_{\gamma} \leq \rho_{\gamma}^{X,Y}(y) \leq C \|y\|_{\gamma} \qquad (3.54)$$

so that

$$S_{\gamma}(X, Y) =: \{y : \rho_{\gamma}^{X,Y}(y) < \infty\} = (X, Y)_{\gamma}.$$

Take now $q, k \in \mathbb{N}$, $m \in \mathbb{N}_*$, $p > 1$, $Y = W_*^{k,\infty}$, and $X = W^{q,2m,p}$. Then, with the notation from (3.5) and (3.6),

$$\rho_{q,k,m,p}(\mu) = \rho_{\gamma}^{X,Y}(\mu), \quad S_{q,k,m,p_p} = S_{\gamma}(X, Y), \qquad (3.55)$$

where $\gamma = (q + k + d/p_*)/2m$. Notice that, in the definition of $\mathcal{S}_{q,k,m,p}$, we do not use exactly $\pi_\gamma(y, (x_n)_n)$, but $\pi_\gamma^{(m)}(y, (x_n)_n)$, defined by

$$\pi_\gamma^{(m)}(y, (x_n)_n) = \sum_{n=1}^{\infty} 2^{n(q+k+d/p_*)} \, \|y - x_n\|_Y + \frac{1}{2^{2mn}} \, \|x_n\|_X$$

$$= \sum_{n=1}^{\infty} 2^{2mn\gamma} \, \|y - x_n\|_Y + \frac{1}{2^{2mn}} \, \|x_n\|_X \,,$$

where $\gamma = (q + k + d/p_*)/2m$. The fact that we use 2^{2mn} instead of 2^n has no impact except for changing the constants in (3.54). So the spaces are the same.

We now turn to a different point. For $p > 1$ and $0 < s < 1$, we denote by $\mathcal{B}^{s,p}$ the Besov space and by $\|f\|_{\mathcal{B}^{s,p}}$ the Besov norm (see Triebel [47] for definitions and notations). Our aim is to give a criterion which guarantees that a function f belongs to $\mathcal{B}^{s,p}$. We will use the classical equality $(W^{1,p}, L^p)_s = \mathcal{B}^{s,p}$.

Lemma 3.6.1. *Let $p > 1$ and $0 < s' < s < 1$. Consider a function $\phi \in C^\infty$ such that $\phi \geq 0$ and $\int \phi = 1$, and let $\phi_\delta(x) = \phi(x/\delta)/\delta^d$ and $\phi_\delta^i(x) = x^i \phi_\delta(x)$. We assume that $f \in L^p$ verifies the following hypotheses: for every $i = 1, \ldots, d$,*

$$\text{(i)} \quad \overline{\lim_{\delta \to 0}} \, \delta^{1-s} \, \|\partial_i(f * \phi_\delta)\|_p < \infty$$

$$\text{(ii)} \quad \overline{\lim_{\delta \to 0}} \, \delta^{-s} \, \|\partial_i(f * \phi_\delta^i)\|_p < \infty. \tag{3.56}$$

Then, $f \in \mathcal{B}^{s',p}$ for every $s' < s$.

Proof. Let $f \in C^1$. We use a Taylor expansion of order one and we obtain

$$f(x) - f * \phi_\varepsilon(x) = \int (f(x) - f(x - y))\phi_\varepsilon(y) dy$$

$$= \int_0^1 d\lambda \int \langle \nabla f(x - \lambda y), y \rangle \, \phi_\varepsilon(y) dy$$

$$= \int_0^1 d\lambda \int \langle \nabla f(x - z), z \rangle \frac{1}{\lambda^d} \phi_\varepsilon\left(\frac{z}{\lambda}\right) \frac{dz}{\lambda^d}$$

$$= \int_0^1 d\lambda \int \langle \nabla f(x - z), z \rangle \, \phi_{\varepsilon\lambda}(z) \frac{dz}{\lambda}$$

$$= \sum_{i=1}^d \int_0^1 \partial_i(f * \phi_{\varepsilon\lambda}^i)(x) \frac{d\lambda}{\lambda}.$$

It follows that

$$\|f - f * \phi_\varepsilon\|_p \leq \sum_{i=1}^d \int_0^1 \|\partial_i(f * \phi_{\varepsilon\lambda}^i)\|_p \frac{d\lambda}{\lambda} \leq d\varepsilon^s \int_0^1 \frac{d\lambda}{\lambda^{1-s}} = C\varepsilon^s.$$

We also have $\|f * \phi_\varepsilon\|_{W^{1,p}} \le C\varepsilon^{-(1-s)}$, so that

$$K(f, \varepsilon) \le \|f - f * \phi_\varepsilon\|_p + \varepsilon \|f * \phi_\varepsilon\|_{W^{1,p}} \le C\varepsilon^s.$$

We conclude that for $s' < s$ we have

$$\int_0^1 \frac{1}{\varepsilon^{s'}} K(f, \varepsilon) \frac{d\varepsilon}{\varepsilon} \le C \int_0^1 \frac{\varepsilon^s}{\varepsilon^{s'}} \frac{d\varepsilon}{\varepsilon} < \infty$$

so $f \in (W^{1,p}, L^p)_{s'} = \mathcal{B}^{s',p}.$ □

3.7 Appendix C: Superkernels

A superkernel $\phi \colon \mathbb{R}^d \to \mathbb{R}$ is a function which belongs to the Schwartz space $\mathcal{S}$ (infinitely differentiable functions with polynomial decay at infinity), $\int \phi(x)dx = 1$, and such that for every non-null multi index $\alpha = (\alpha_1, \dots, \alpha_d) \in \mathbb{N}^d$ we have

$$\int y^\alpha \phi(y)dy = 0, \quad y^\alpha = \prod_{i=1}^d y_i^{\alpha_i}. \tag{3.57}$$

See Remark 1 in [29, Section 3] for the construction of a superkernel. The corresponding ϕ_δ, $\delta \in (0,1)$, is defined by

$$\phi_\delta(y) = \frac{1}{\delta^d} \phi\left(\frac{y}{\delta}\right).$$

For a function f we denote $f_\delta = f * \phi_\delta$. We will work with the norms $\|f\|_{k,\infty}$ and $\|f\|_{q,l,p}$ defined in (3.1) and in (3.2). We have

Lemma 3.7.1. (i) *Let $k, q \in \mathbb{N}$, $l > d$, and $p > 1$. There exists a universal constant C such that, for every $f \in W^{q,l,p}$, we have*

$$\|f - f_\delta\|_{W_*^{k,\infty}} \le C \|f\|_{q,l,p} \delta^{q+k}. \tag{3.58}$$

(ii) *Let $l > d$, $n, q \in \mathbb{N}$, with $n \ge q$, and let $p > 1$. There exists a universal constant C such that*

$$\|f_\delta\|_{n,l,p} \le C \|f\|_{q,l,p} \delta^{-(n-q)}. \tag{3.59}$$

Proof. (i) Without loss of generality, we may suppose that $f \in C_b^\infty$. Using a Taylor expansion of order $q + k$,

$$f(x) - f_\delta(x) = \int (f(x) - f(y))\phi_\delta(x - y)dy$$

$$= \int I(x, y)\phi_\delta(x - y)dy + \int R(x, y)\phi_\delta(x - y)dy$$

with

$$I(x,y) = \sum_{i=1}^{q+k-1} \frac{1}{i!} \sum_{|\alpha|=i} \partial_\alpha f(x)(x-y)^\alpha,$$

$$R(x,y) = \frac{1}{(q+k)!} \sum_{|\alpha|=q+k} \int_0^1 \partial_\alpha f(x+\lambda(y-x))(x-y)^\alpha d\lambda.$$

Using (3.57), we obtain $\int I(x,y)\phi_\delta(x-y)dy = 0$ and, by a change of variable, we get

$$\int R(x,y)\phi_\delta(x-y)dy = \frac{1}{(q+k)!} \sum_{|\alpha|=q+k} \int_0^1 \int dz \phi_\delta(z)\partial_\alpha f(x+\lambda z)z^\alpha d\lambda.$$

We consider now $g \in W^{k,\infty}$ and write

$$\int (f(x)-f_\delta(x))g(x)dx = \frac{1}{(q+k)!} \sum_{|\alpha|=q+k} \int_0^1 d\lambda \int dz \phi_\delta(z)z^\alpha \int \partial^\alpha f(x+\lambda z)g(x)dx.$$

We write $\partial_\gamma f(x) = u_l(x)v_\gamma(x)$ with $u_l(x) = (1+|x|^2)^{-l/2}$ and $v_\gamma(x) = (1+|x|^2)^{l/2}\partial_\gamma f(x)$. Notice that for every $l > d$, $\|u_l\|_{p_*} < \infty$. Then using Hölder's inequality we have

$$\int |\partial_\gamma f(x)|\, dx \le C\, \|u_l\|_{p_*}\, \|v_\gamma\|_p \le C\, \|u_l\|_{p_*}\, \|f\|_{q,l,p}\,,$$

and so

$$\left| \int_0^1 \int dz \phi_\delta(z)z^\alpha \int \partial_\alpha f(x+\lambda z)g(x)dxd\lambda \right| \le C\, \|f\|_{q,l,p}\, \|g\|_{k,\infty} \int \phi_\delta(z)\, |z|^{k+q}\, dz$$

$$\le C\, \|f\|_{q,l,p}\, \|g\|_{k,\infty}\, \delta^{k+q}.$$

(ii) Let α be a multiindex with $|\alpha| = n$ and let β, γ be a splitting of α with $|\beta| = q$ and $|\gamma| = n - q$. Using the triangle inequality, for every y we have $1 + |x| \le (1+|y|)(1+|x-y|)$. Then,

$$u(x) := (1+|x|)^l\, |\partial_\alpha f_\delta(x)| = (1+|x|)^l\, |\partial_\beta f * \partial_\gamma \phi_\delta(x)|$$

$$\le \int (1+|x|)^l\, |\partial_\beta f(y)|\, |\partial_\gamma \phi_\delta(x-y)|\, dy \le \alpha * \beta(x),$$

with

$$\alpha(y) = (1+|y|)^l\, |\partial^\beta f(y)|\,, \quad \beta(z) = (1+|z|)^l\, |\partial^\gamma \phi_\delta(z)|\,.$$

Using (3.47) we obtain

$$\|u\|_p \le \|\alpha * \beta\|_p \le \|\beta\|_1\, \|\alpha\|_p \le \frac{C}{\delta^{n-q}}\, \|\alpha\|_p = \frac{C}{\delta^{n-q}}\, \|f_{\beta,l}\|_p\,. \qquad \square$$

Bibliography

[1] V. Bally, *An elementary introduction to Malliavin calculus*. Rapport de recherche 4718, INRIA 2003.

[2] V. Bally and L. Caramellino, *Riesz transform and integration by parts formulas for random variables*. Stoch. Proc. Appl. **121** (2011), 1332–1355.

[3] V. Bally and L. Caramellino, *Positivity and lower bounds for the density of Wiener functionals*. Potential Analysis **39** (2013), 141–168.

[4] V. Bally and L. Caramellino, *On the distances between probability density functions*. Electron. J. Probab. **19** (2014), no. 110, 1–33.

[5] V. Bally and L. Caramellino, *Convergence and regularity of probability laws by using an interpolation method*. Annals of Probability, to appear (accepted on November 2015). arXiv:1409.3118 (past version).

[6] V. Bally and L. Caramellino, *Regularity of Wiener functionals under a Hörmander type condition of order one*. Preprint (2013), arXiv:1307.3942.

[7] V. Bally and L. Caramellino, *Asymptotic development for the CLT in total variation distance*. Preprint (2014), arXiv:1407.0896.

[8] V. Bally and E. Clément, *Integration by parts formula and applications to equations with jumps*. Prob. Theory Related Fields **151** (2011), 613–657.

[9] V. Bally and E. Clément, *Integration by parts formula with respect to jump times for stochastic differential equations*. In: Stochastic Analysis 2010, 7–29, Springer, Heidelberg, 2011.

[10] V. Bally and N. Fournier, *Regularization properties od the 2D homogeneous Boltzmann equation without cutoff*. Prob. Theory Related Fields **151** (2011), 659–704.

[11] V. Bally and E. Pardoux, *Malliavin calculus for white noise driven parabolic SPDEs*. Potential Analysis **9** (1998), 27–64.

[12] V. Bally and D. Talay, *The law of the Euler scheme for stochastic differential equations. I. Convergence rate of the distribution function*. Prob. Theory Related Fields **104** (1996), 43–60.

[13] V. Bally and D. Talay, *The law of the Euler scheme for stochastic differential equations. II. Convergence rate of the density*. Monte Carlo Methods Appl. **2** (1996), 93–128.

[14] G. Ben Arous and R. Léandre, *Décroissance exponentielle du noyau de la chaleur sur la diagonale. II*. Prob. Theory Related Fields **90** (1991), 377–402.

[15] R.N. Bhattacharaya and R. Ranga Rao, *Normal Approximation and Asymptotic Expansions.* SIAM Classics in Applied Mathematics, 2010.

[16] K. Bichtler, J.B. Gravereaux and J. Jacod, *Malliavin Calculus for Processes with Jumps.* Gordon and Breach Science Publishers, 1987.

[17] J.M. Bismut, *Calcul des variations stochastique et processus de sauts.* Z. Wahrscheinlichkeitstheorie und Verw. Gebiete **63** (1983), 147–235.

[18] H. Brezis, *Analyse Fonctionelle. Théorie et Applications.* Masson, Paris, 1983.

[19] A. Debussche and N. Fournier, *Existence of densities for stable-like driven SDEs with Hölder continuous coefficients.* J. Funct. Anal. **264** (2013), 1757–1778.

[20] A. Debussche and M. Romito, *Existence of densities for the 3D Navier–Stokes equations driven by Gaussian noise.* Prob. Theory Related Fields **158** (2014), 575–596.

[21] S. De Marco, *Smoothness and asymptotic estimates of densities for SDEs with locally smooth coefficients and applications to square-root diffusions.* Ann. Appl. Probab. **21** (2011), 1282–1321.

[22] G. Di Nunno, B. Øksendal and F. Proske, *Malliavin Calculus for Lévy Processes with Applications to Finance.* Universitext. Springer-Verlag, Berlin, 2009.

[23] J. Dziubanski, *Triebel–Lizorkin spaces associated with Laguerre and Hermite expansions.* Proc. Amer. Math. Soc. **125** (1997), 3547–3554.

[24] J. Epperson, *Hermite and Laguerre wave packet expansions.* Studia Math. **126** (1997), 199–217.

[25] N. Fournier, *Finiteness of entropy for the homogeneous Boltzmann equation with measure initial condition.* Ann. Appl. Prob. **25** (2015), no. 2, 860–897.

[26] N. Fournier and J. Printems, *Absolute continuity of some one-dimensional processes.* Bernoulli **16** (2010), 343–360.

[27] E. Gobet and C. Labart, *Sharp estimates for the convergence of the density of the Euler scheme in small time.* Electron. Commun. Probab. **13** (2008), 352–363.

[28] J. Guyon, *Euler scheme and tempered distributions.* Stoch. Proc. Appl. **116** (2006), 877–904.

[29] A. Kebaier and A. Kohatsu-Higa, *An optimal control variance reduction method for density estimation.* Stoch. Proc. Appl. **118** (2008), 2143–2180.

[30] A. Kohatsu-Higa and A. Makhlouf, *Estimates for the density of functionals of SDEs with irregular drift.* Stochastic Process. Appl. **123** (2013), 1716–1728.

[31] A. Kohatsu-Higa and K. Yasuda, *Estimating multidimensional density functions for random variables in Wiener space.* C. R. Math. Acad. Sci. Paris **346** (2008), 335–338.

[32] A. Kohatsu-Higa and K. Yasuda, *Estimating multidimensional density functions using the Malliavin–Thalmaier formula.* SIAM J. Numer. Anal. **47** (2009), 1546–1575.

[33] P. Malliavin, *Stochastic calculus of variations and hypoelliptic operators.* In: Proc. Int. Symp. Stoch. Diff. Equations, Kyoto, 1976. Wiley (1978), 195–263.

[34] P. Malliavin, *Stochastic Analysis.* Springer, 1997.

[35] P. Malliavin and E. Nualart, *Density minoration of a strongly non degenerated random variable.* J. Funct. Anal. **256** (2009), 4197–4214.

[36] P. Malliavin and A. Thalmaier, *Stochastic Calculus of Variations in Mathematical Finance.* Springer-Verlag, Berlin, 2006.

[37] S. Ninomiya and N. Victoir, *Weak approximation of stochastic differential equations and applications to derivative pricing.* Appl. Math. Finance **15** (2008), 107–121.

[38] I. Nourdin, D. Nualart and G. Poly, *Absolute continuity and convergence of densities for random vectors on Wiener chaos.* Electron. J. Probab. **18** (2013), 1–19.

[39] I. Nourdin and G. Peccati, *Normal Approximations with Malliavin Calculus. From Stein's Method to Universality.* Cambridge Tracts in Mathematics **192**. Cambridge University Press, 2012.

[40] I. Nourdin and G. Poly, *Convergence in total variation on Wiener chaos.* Stoch. Proc. Appl. **123** (2013), 651–674.

[41] D. Nualart, *The Malliavin Calculus and Related Topics.* Second edition. Springer-Verlag, Berlin, 2006.

[42] E. Pardoux and T. Zhang, *Absolute continuity for the law of the solution of a parabolic SPDE.* J. Funct. Anal. **112** (1993), 447–458.

[43] P. Petrushev and Y. Xu, *Decomposition of spaces of distributions induced by Hermite expansions.* J. Fourier Anal. and Appl. **14** (2008), 372–414.

[44] M. Sanz-Solé, *Malliavin Calculus, with Applications to Stochastic Partial Differential Equations.* EPFL Press, 2005.

[45] I. Shigekawa, *Stochastic Analysis.* Translations of Mathematical Monographs, 224. Iwanami Series in Modern Mathematics. American Mathematical Society, Providence, RI, 2004.

[46] D. Talay and L. Tubaro, *Expansion of the global error for numerical schemes solving stochastic differential equations.* Stochastic Anal. Appl. **8** (1990), 94–120.

[47] H. Triebel, *Theory of Function Spaces III*. Birkhäuser Verlag, 2006.

[48] J.B. Walsh, *An introduction to stochastic partial differential equations.* In: École d'Eté de Probabilités de Saint Flour XV, Lecture Notes in Math. 1180, Springer 1986, 226–437.

Part II

Functional Itô Calculus and Functional Kolmogorov Equations

Rama Cont

Preface

The *Functional Itô Calculus* is a non-anticipative functional calculus which extends the Itô calculus to path-dependent functionals of stochastic processes. We present an overview of the foundations and key results of this calculus, first using a pathwise approach, based on the notion of directional derivative introduced by Dupire, then using a probabilistic approach, which leads to a weak, Sobolev type, functional calculus for square-integrable functionals of semimartingales, without any smoothness assumption on the functionals. The latter construction is shown to have connections with the Malliavin Calculus and leads to computable martingale representation formulae. Finally, we show how the Functional Itô Calculus may be used to extend the connections between diffusion processes and parabolic equations to a non-Markovian, path-dependent setting: this leads to a new class of 'path-dependent' partial differential equations, namely the so-called *functional Kolmogorov equations*, of which we study the key properties.

These notes are based on a series of lectures given at various summer schools: Tokyo (Tokyo Metropolitan University, July 2010), Munich (Ludwig Maximilians University, October 2010), the Spring School "Stochastic Models in Finance and Insurance" (Jena, 2011), the International Congress for Industrial and Applied Mathematics (Vancouver, 2011), and the "Barcelona Summer School on Stochastic Analysis" (Barcelona, July 2012).

I am grateful to Hans Engelbert, Jean Jacod, Hans Föllmer, Yuri Kabanov, Arman Khaledian, Shigeo Kusuoka, Bernt Oksendal, Candia Riga, Josep Vives, and especially the late Paul Malliavin for helpful comments and discussions.

Rama Cont

Chapter 4

Overview

4.1 Functional Itô Calculus

Many questions in stochastic analysis and its applications in statistics of processes, physics, or mathematical finance involve *path-dependent functionals* of stochastic processes and there has been a sustained interest in developing an analytical framework for the systematic study of such path-dependent functionals.

When the underlying stochastic process is the Wiener process, the Malliavin Calculus [4, 52, 55, 56, 67, 73] has proven to be a powerful tool for investigating various properties of Wiener functionals. The Malliavin Calculus, which is as a weak functional calculus on Wiener space, leads to differential representations of Wiener functionals in terms of *anticipative* processes [5, 37, 56]. However, the interpretation and computability of such anticipative quantities poses some challenges, especially in applications such as mathematical physics or optimal control, where causality, or non-anticipativeness, is a key constraint.

In a recent insightful work, motivated by applications in mathematical finance, Dupire [21] has proposed a method for defining a non-anticipative calculus which extends the Itô Calculus to path-dependent functionals of stochastic processes. The idea can be intuitively explained by first considering the variations of a functional along a *piecewise constant* path. Any (right continuous) piecewise constant path, represented as

$$\omega(t) = \sum_{k=1}^{n} x_k 1_{[t_k, t_{k+1}[}(t),$$

is simply a finite sequence of 'horizonal' and 'vertical' moves, so the variation of a (time dependent) functional $F(t, \omega)$ along such a path ω is composed of

(i) 'horizontal increments': variations of $F(t, \omega)$ between each time point t_i and the next, and

(ii) 'vertical increments': variations of $F(t, \omega)$ at each discontinuity point of ω.

If one can control the behavior of F under these two types of path perturbations, then one can reconstitute its variations along any piecewise constant path. Under additional continuity assumptions, this control can be extended to any càdlàg path using a density argument.

This intuition was formalized by Dupire [21] by introducing directional derivatives corresponding to infinitesimal versions of these variations: given a (time dependent) functional $F \colon [0, T] \times D([0, T], \mathbb{R}) \to \mathbb{R}$ defined on the space $D([0, T], \mathbb{R})$

of right continuous paths with left limits, Dupire introduced a directional derivative which quantifies the sensitivity of the functional to a shift in the future portion of the underlying path $\omega \in D([0,T], \mathbb{R})$:

$$\nabla_\omega F(t, \omega) = \lim_{\epsilon \to 0} \frac{F(t, \omega + \epsilon 1_{[t,T]}) - F(t, \omega)}{\epsilon},$$

as well as a time derivative corresponding to the sensitivity of F to a small 'horizontal extension' of the path. Since any càdlàg path may be approximated, in supremum norm, by piecewise constant paths, this suggests that one may control the functional F on the entire space $D([0,T], \mathbb{R})$ if F is twice differentiable in the above sense and $F, \nabla_\omega F, \nabla_\omega^2 F$ are continuous in supremum norm; under these assumptions, one can then obtain a change of variable formula for $F(X)$ for any Itô process X.

As this brief description already suggests, the essence of this approach is pathwise. While Dupire's original presentation [21] uses probabilistic arguments and Itô Calculus, one can in fact do it entirely without such arguments and derive these results in a purely analytical framework without any reference to probability. This task, undertaken in [8, 9] and continued in [6], leads to a pathwise functional calculus for *non-anticipative functionals* which clearly identifies the set of paths to which the calculus is applicable. The pathwise nature of all quantities involved makes this approach quite intuitive and appealing for applications, especially in finance; see Cont and Riga [13].

However, once a probability measure is introduced on the space of paths, under which the canonical process is a semimartingale, one can go much further: the introduction of a reference measure allows to consider quantities which are defined almost everywhere and construct a *weak functional calculus* for stochastic processes defined on the canonical filtration. Unlike the pathwise theory, this construction, developed in [7, 11], is applicable to all square-integrable functionals *without any regularity condition*. This calculus can be seen as a non-anticipative analog of the Malliavin Calculus.

The Functional Itô Calculus has led to various applications in the study of path dependent functionals of stochastic processes. Here, we focus on two particular directions: martingale representation formulas and functional ('path dependent') Kolmogorov equations [10].

4.2   Martingale representation formulas

The representation of martingales as stochastic integrals is an important result in stochastic analysis, with many applications in control theory and mathematical finance. One of the challenges in this regard has been to obtain explicit versions of such representations, which may then be used to compute or simulate such martingale representations. The well-known Clark–Haussmann–Ocone formula [55, 56], which expresses the martingale representation theorem in terms of the Malliavin derivative, is one such tool and has inspired various algorithms for the simulation of such representations, see [33].

One of the applications of the Functional Itô Calculus is to derive explicit, computable versions of such martingale representation formulas, without resorting to the anticipative quantities such as the Malliavin derivative. This approach, developed in [7, 11], leads to simple algorithms for computing martingale representations which have straightforward interpretations in terms of sensitivity analysis [12].

4.3 Functional Kolmogorov equations and path dependent PDEs

One of the important applications of the Itô Calculus has been to characterize the deep link between Markov processes and partial differential equations of parabolic type [2]. A pillar of this approach is the analytical characterization of a Markov process by Kolmogorov's backward and forward equations [46]. These equations have led to many developments in the theory of Markov processes and stochastic control theory, including the theory of controlled Markov processes and their links with viscosity solutions of PDEs [29].

The Functional Itô Calculus provides a natural setting for extending many of these results to more general, non-Markovian semimartingales, leading to a new class of partial differential equations on path space —*functional Kolmogorov equations*— which have only started to be explored [10, 14, 22]. This class of PDEs on the space of continuous functions is distinct from the infinite-dimensional Kolmogorov equations studied in the literature [15]. Functional Kolmogorov equations have many properties in common with their finite-dimensional counterparts and lead to new Feynman–Kac formulas for path dependent functionals of semimartingales [10]. We will explore this topic in Chapter 8. Extensions of these connections to the fully nonlinear case and their connections to non-Markovian stochastic control and forward-backward stochastic differential equations (FBSDEs) currently constitute an active research topic, see for exampe [10, 14, 22, 23, 60].

4.4 Outline

These notes, based on lectures given at the Barcelona Summer School on Stochastic Analysis (2012), constitute an introduction to the foundations and applications of the Functional Itô Calculus.

- We first develop a *pathwise calculus for non-anticipative functionals* possessing some directional derivatives, by combining Dupire's idea with insights from the early work of Hans Föllmer [31]. This construction is purely analytical (i.e., non-probabilistic) and applicable to functionals of paths with finite quadratic variation. Applied to functionals of a semimartingale, it yields a functional extension of the Itô formula applicable to functionals which are continuous in the supremum norm and admit certain directional derivatives. This construction and its various extensions, which are based on [8, 9, 21] are described in Chapters 5 and 6. As a by-product, we obtain a method for constructing pathwise integrals with respect to paths of infinite variation but finite quadratic variation, for a class of integrands which may be described as 'vertical 1-forms'; the connection between this pathwise integral and 'rough path' theory is described in Subsection 5.3.3.

- In Chapter 7 we extend this pathwise calculus to a *'weak' functional calculus* applicable to square-integrable adapted functionals with *no regularity condition* on the path dependence. This construction uses the probabilistic structure of the Itô integral to construct an extension of Dupire's derivative operator to all square-integrable semimartingales and introduce Sobolev spaces of non-anticipative functionals to which weak versions of the functional Itô formula applies. The resulting operator is a weak functional derivative which may be regarded as a *non-anticipative* counterpart of the Malliavin derivative (Section 7.4). This construction, which extends the applicability of the Functional Itô Calculus to a large class of functionals, is based on [11]. The relation with the Malliavin derivative is described in Section 7.4. One of the applications of this construction is to obtain explicit and computable integral representation formulas for martingales (Section 7.2 and Theorem 7.3.4).

- Chapter 8 uses the Functional Itô Calculus to introduce *Functional Kolmogorov equations*, a new class of partial differential equations on the space of continuous functions which extend the classical backward Kolmogorov PDE to processes with path dependent characteristics. We first present some key properties of classical solutions for this class of equations, and their relation with FBSDEs with path dependent coefficients (Section 8.2) and non-Markovian stochastic control problems (Section 8.3). Finally, in Section 8.4 we introduce a notion of weak solution for the functional Kolmogorov equation and characterize square-integrable martingales as weak solutions of the functional Kolmogorov equation.

Notations

For the rest of the manuscript, we denote by

- S_d^+, the set of symmetric positive $d \times d$ matrices with real entries;
- $D([0,T], \mathbb{R}^d)$, the space of functions on $[0,T]$ with values in $\mathbb{R}^d$ which are right continuous functions with left limits (càdlàg); and
- $C^0([0,T], \mathbb{R}^d)$, the space of continuous functions on $[0,T]$ with values in $\mathbb{R}^d$.

Both spaces above are equipped with the supremum norm, denoted $\| \cdot \|_\infty$. We further denote by

- $C^k(\mathbb{R}^d)$, the space of k-times continuously differentiable real-valued functions on $\mathbb{R}^d$;
- $H^1([0,T], \mathbb{R})$, the Sobolev space of real-valued absolutely continuous functions on $[0,T]$ whose Radon–Nikodým derivative with respect to the Lebesgue measure is square-integrable.

For a path $\omega \in D([0,T], \mathbb{R}^d)$, we denote by

- $\omega(t-) = \lim_{s \to t, s < t} \omega(s)$, its left limit at t;
- $\Delta\omega(t) = \omega(t) - \omega(t-)$, its discontinuity at t;
- $\|\omega\|_\infty = \sup\{|\omega(t)|, t \in [0,T]\}$;
- $\omega(t) \in \mathbb{R}^d$, the value of ω at t;
- $\omega_t = \omega(t \wedge \cdot)$, the path stopped at t; and
- $\omega_{t-} = \omega \, 1_{[0,t[} + \omega(t-) \, 1_{[t,T]}$.

Note that $\omega_{t-} \in D([0,T], \mathbb{R}^d)$ is càdlàg and should *not* be confused with the càglàd path $u \mapsto \omega(u-)$. For a càdlàg stochastic process X we similarly denote by

- $X(t)$, its value;
- $X_t = (X(u \wedge t), 0 \le u \le T)$, the process stopped at t; and
- $X_{t-}(u) = X(u) \, 1_{[0,t[}(u) + X(t-) \, 1_{[t,T]}(u)$.

For general definitions and concepts related to stochastic processes, we refer to the treatises by Dellacherie and Meyer [19] and by Protter [62].

Chapter 5

Pathwise calculus for non-anticipative functionals

The focus of these lectures is to define a calculus which can be used to describe the variations of interesting classes of functionals of a given reference stochastic process X. In order to cover interesting examples of processes, we allow X to have right-continuous paths with left limits, i.e., its paths lie in the space $D([0,T], \mathbb{R}^d)$ of càdlàg paths. In order to include the important case of Brownian diffusion and diffusion processes, we allow these paths to have infinite variation. It is then known that the results of Newtonian calculus and Riemann–Stieltjes integration do not apply to the paths of such processes. Itô's stochastic calculus [19, 40, 41, 53, 62] provides a way out by limiting the class of integrands to *non-anticipative*, or adapted processes; this concept plays an important role in what follows.

Although the classical framework of Itô Calculus is developed in a probabilistic setting, Föllmer [31] was the first to point out that many of the ingredients at the core of this theory —in particular the Itô formula— may in fact be constructed pathwise. Föllmer [31] further identified the concept of *finite quadratic variation* as the relevant property of the path needed to derive the Itô formula.

In this first part, we combine the insights from Föllmer [31] with the ideas of Dupire [21] to construct a pathwise functional calculus for non-anticipative functionals defined on the space $\Omega = D([0,T], \mathbb{R}^d)$ of càdlàg paths. We first introduce the notion of *non-anticipative*, or causal, functional (Section 5.1) and show how these functionals naturally arise in the representation of processes adapted to the natural filtration of a given reference process. We then introduce, following Dupire [21], the directional derivatives which form the basis of this calculus: the horizontal and the vertical derivatives, introduced in Section 5.2. The introduction of these particular directional derivatives, unnatural at first sight, is justified a posteriori by the functional change of variable formula (Section 5.3), which shows that the horizonal and vertical derivatives are precisely the quantities needed to describe the variations of a functional along a càdlàg path with finite quadratic variation.

The results in this chapter are entirely 'pathwise' and do not make use of any probability measure. In Section 5.5, we identify important classes of stochastic processes for which these results apply almost surely.

5.1   Non-anticipative functionals

Let X be the canonical process on $\Omega = D([0,T], \mathbb{R}^d)$, and let $\mathbb{F}^0 = (\mathcal{F}_t^0)_{t \in [0,T]}$ be the filtration generated by X.

A process Y on $(\Omega, \mathcal{F}_T^0)$ adapted to $\mathbb{F}^0$ may be represented as a family of functionals $Y(t, \cdot) \colon \Omega \to \mathbb{R}$ with the property that $Y(t, \cdot)$ only depends on the path stopped at t,

$$Y(t, \omega) = Y(t, \omega(. \wedge t)),$$

so one can represent Y as

$$Y(t, \omega) = F(t, \omega_t) \quad \text{for some functional} \quad F \colon [0, T] \times D([0, T], \mathbb{R}^d) \longrightarrow \mathbb{R},$$

where $F(t, \cdot)$ only needs to be defined on the set of paths stopped at t.

This motivates us to view adapted processes as functionals on the space of *stopped paths*: a stopped path is an equivalence class in $[0, T] \times D([0, T], \mathbb{R}^d)$ for the following equivalence relation:

$$(t, \omega) \sim (t', \omega') \iff (t = t' \quad \text{and} \quad \omega_t = \omega_{t'}), \tag{5.1}$$

where $\omega_t = \omega(t \wedge \cdot)$.

The space of stopped paths can be defined as the quotient space of $[0, T] \times D([0, T], \mathbb{R}^d)$ by the equivalence relation (5.1):

$$\Lambda_T^d = \{(t, \omega_t), \ (t, \omega) \in [0, T] \times D([0, T], \mathbb{R}^d)\} = [0, T] \times D([0, T], \mathbb{R}^d) \, / \sim .$$

We endow this set with a metric space structure by defining the distance:

$$d_\infty((t, \omega), (t', \omega')) = \sup_{u \in [0, T]} |\omega(u \wedge t) - \omega'(u \wedge t')| + |t - t'|.$$

$$= \|\omega_t - \omega'_{t'}\|_\infty + |t - t'|.$$

Then, (Λ_T^d, d_∞) is a complete metric space and the set of continuous stopped paths

$$\mathcal{W}_T^d = \{(t, \omega) \in \Lambda_T^d, \ \omega \in C^0([0, T], \mathbb{R}^d)\}$$

is a closed subset of (Λ_T^d, d_∞). When the context is clear we will drop the superscript d, and denote these spaces Λ_T and $\mathcal{W}_T$, respectively.

We now define a non-anticipative functional as a measurable map on the space (Λ_T, d_∞) of stopped paths [8]:

Definition 5.1.1 (Non-anticipative (causal) functional). *A non-anticipative functional on $D([0, T], \mathbb{R}^d)$ is a measurable map $F \colon (\Lambda_T, d_\infty) \to \mathbb{R}$ on the space of stopped paths, (Λ_T, d_∞).*

This notion of causality is natural when dealing with physical phenomena as well as in control theory [30].

A non-anticipative functional may also be seen as a family $F = (F_t)_{t\in[0,T]}$ of $\mathcal{F}_t^0$-measurable maps $F_t \colon (D([0,t],\mathbb{R}^d), \|.\|_\infty) \to \mathbb{R}$. Definition 5.1.1 amounts to requiring joint measurability of these maps in (t,ω).

One can alternatively represent F as a map

$$F \colon \bigcup_{t\in[0,T]} D([0,t],\mathbb{R}^d) \longrightarrow \mathbb{R}$$

on the vector bundle $\bigcup_{t\in[0,T]} D([0,t],\mathbb{R}^d)$. This is the original point of view developed in [8, 9, 21], but leads to slightly more complicated notations. We will follow here Definition 5.1.1, which has the advantage of dealing with paths defined on a fixed interval and alleviating notations.

Any progressively measurable process Y defined on the filtered canonical space $(\Omega, (\mathcal{F}_t^0)_{t\in[0,T]})$ may, in fact, be represented by such a non-anticipative functional F:

$$Y(t) = F(t, X(t \wedge \cdot)) = F(t, X_t);$$

see [19, Vol. I]. We will write $Y = F(X)$. Conversely, any non-anticipative functional F applied to X yields a progressively measurable process $Y(t) = F(t, X_t)$ adapted to the filtration $\mathcal{F}_t^0$.

We now define the notion of *predictable* functional as a non-anticipative functional whose value depends only on the past, but not the present value, of the path. Recall the notation:

$$\omega_{t-} = \omega\, 1_{[0,t[} + \omega(t-)1_{[t,T]}.$$

Definition 5.1.2 (Predictable functional). *A non-anticipative functional*

$$F \colon (\Lambda_T, d_\infty) \longrightarrow \mathbb{R}$$

is called predictable if $F(t,\omega) = F(t, \omega_{t-})$ *for every* $(t,\omega) \in \Lambda_T$.

This terminology is justified by the following property: if X is a càdlàg and $\mathcal{F}_t$-adapted process and F is a predictable functional, then $Y(t) = F(t, X_t)$ defines an $\mathcal{F}_t$-predictable process.

Having equipped the space of stopped paths with the metric d_∞, we can now define various notions of continuity for non-anticipative functionals.

Definition 5.1.3 (Joint continuity in (t,ω)). *A continuous non-anticipative functional is a continuous map* $F \colon (\Lambda_T, d_\infty) \to \mathbb{R}$ *such that:* $\forall(t,\omega) \in \Lambda_T$,

$$\forall \epsilon > 0, \exists \eta > 0, \forall(t',\omega') \in \Lambda_T, d_\infty((t,\omega),(t',\omega')) < \eta \Rightarrow |F(t,\omega) - F(t',\omega')| < \epsilon.$$

The set of continuous non-anticipative functionals is denoted by $\mathbb{C}^{0,0}(\Lambda_T)$.

A non-anticipative functional F is said to be *continuous at fixed times* if for all $t \in [0,T[$, the map

$$F(t,.) \colon (D([0,T],\mathbb{R}^d), \|.\|_\infty) \longrightarrow \mathbb{R}$$

is continuous.

The following notion, which we will use most, distinguishes the time variable and is more adapted to probabilistic applications.

Definition 5.1.4 (Left-continuous non-anticipative functionals). *Define* $\mathbb{C}_l^{0,0}(\Lambda_T)$ *as the set of non-anticipative functionals F which are continuous at fixed times, and which satisfy:* $\forall (t,\omega) \in \Lambda_T,\ \forall \epsilon > 0,\ \exists \eta > 0,\ \forall (t',\omega') \in \Lambda_T,$

$$t' < t \quad \text{and} \quad d_\infty((t,\omega),(t',\omega')\,) < \eta \Longrightarrow |F(t,\omega) - F(t',\omega')| < \epsilon.$$

The image of a left-continuous path by a left-continuous functional is again left-continuous: $\forall F \in \mathbb{C}_l^{0,0}(\Lambda_T), \forall \omega \in D([0,T],\mathbb{R}^d),\ t \mapsto F(t,\omega_{t-})$ is left-continuous.

Let U be a càdlàg $\mathcal{F}_t^0$-adapted process. A non-anticipative functional F applied to U generates a process adapted to the natural filtration $\mathcal{F}_t^U$ of U:

$$Y(t) = F(t, \{U(s), 0 \leq s \leq t\}) = F(t, U_t).$$

The following result is shown in [8, Proposition 1]:

Proposition 5.1.5. *Let $F \in \mathbb{C}_l^{0,0}(\Lambda_T)$ and U be a càdlàg $\mathcal{F}_t^0$-adapted process. Then,*

(i) $Y(t) = F(t, U_{t-})$ *is a left-continuous $\mathcal{F}_t^0$-adapted process;*

(ii) $Z \colon [0,T] \times \Omega \to F(t, U_t(\omega))$ *is an optional process;*

(iii) *if $F \in \mathbb{C}^{0,0}(\Lambda_T)$, then $Z(t) = F(t, U_t)$ is a càdlàg process, continuous at all continuity points of U.*

We also introduce the notion of 'local boundedness' for functionals: we call a functional F "boundedness preserving" if it is bounded on each bounded set of paths:

Definition 5.1.6 (Boundedness preserving functionals). *Define $\mathbb{B}(\Lambda_T)$ as the set of non-anticipative functionals $F \colon \Lambda_T \to \mathbb{R}$ such that, for any compact $K \subset \mathbb{R}^d$ and $t_0 < T$,*

$$\exists C_{K,t_0} > 0,\ \forall t \leq t_0,\ \forall \omega \in D([0,T], \mathbb{R}^d), \quad \omega([0,t]) \subset K \Longrightarrow |F(t,\omega)| \leq C_{K,t_0}. \tag{5.2}$$

5.2 Horizontal and vertical derivatives

To understand the key idea behind this pathwise calculus, consider first the case of a non-anticipative functional F applied to a piecewise constant path

$$\omega = \sum_{k=1}^n x_k 1_{[t_k, t_{k+1}[} \in D([0,T], \mathbb{R}^d).$$

Any such piecewise constant path ω is obtained by a finite sequence of operations consisting of

- "horizontal stretching" of the path from t_k to t_{k+1}, followed by

- the addition of a jump at each discontinuity point.

In terms of the stopped path (t, ω), these two operations correspond to

- incrementing the first component: $(t_k, \omega_{t_k}) \to (t_{k+1}, \omega_{t_k})$, and

- shifting the path by $(x_{k+1} - x_k) 1_{[t_{k+1}, T]}$:

$$\omega_{t_{k+1}} = \omega_{t_k} + (x_{k+1} - x_k) 1_{[t_{k+1}, T]}.$$

The variation of a non-anticipative functional along ω can also be decomposed into the corresponding 'horizontal' and 'vertical' increments:

$$F(t_{k+1}, \omega_{t_{k+1}}) - F(t_k, \omega_{t_k}) =$$

$$\underbrace{F(t_{k+1}, \omega_{t_{k+1}}) - F(t_{k+1}, \omega_{t_k})}_{\text{vertical increment}} + \underbrace{F(t_{k+1}, \omega_{t_k}) - F(t_k, \omega_{t_k})}_{\text{horizontal increment}}.$$

Thus, if one can control the behavior of F under these two types of path perturbations, then one can compute the variations of F along any piecewise constant path ω. If, furthermore, these operations may be controlled with respect to the supremum norm, then one can use a density argument to extend this construction to all càdlàg paths.

Dupire [21] formalized this idea by introducing directional derivatives corresponding to infinitesimal versions of these variations: the *horizontal* and *vertical* derivatives, which we now define.

5.2.1 Horizontal derivative

Let us introduce the notion of *horizontal extension* of a stopped path (t, ω_t) to $[0, t + h]$: this is simply the stopped path $(t + h, \omega_t)$. Denoting $\omega_t = \omega(t \wedge \cdot)$ recall that, for any non-anticipative functional,

$$\forall (t, \omega) \in [0, T] \times D([0, T], \mathbb{R}^d), \quad F(t, \omega) = F(t, \omega_t).$$

Definition 5.2.1 (Horizontal derivative). *A non-anticipative functional $F \colon \Lambda_T \to \mathbb{R}$ is said to be horizontally differentiable at $(t, \omega) \in \Lambda_T$ if the limit*

$$\mathcal{D}F\,(t, \omega) = \lim_{h \to 0^+} \frac{F(t + h, \omega_t) - F(t, \omega_t)}{h} \qquad exists. \tag{5.3}$$

We call $\mathcal{D}F\,(t, \omega)$ the horizontal derivative $\mathcal{D}F$ of F at (t, ω).

Importantly, note that given the non-anticipative nature of F, the first term in the numerator of (5.3) depends on $\omega_t = \omega(t \wedge \cdot)$, not ω_{t+h}. If F is horizontally differentiable at all $(t, \omega) \in \Lambda_T$, then the map $\mathcal{D}F \colon (t, \omega) \mapsto \mathcal{D}F\,(t, \omega)$ defines a non-anticipative functional which is $\mathcal{F}_t^0$-measurable, without any assumption on the right-continuity of the filtration.

If $F(t,\omega) = f(t,\omega(t))$ with $f \in C^{1,1}([0,T] \times \mathbb{R}^d)$, then $\mathcal{D}F(t,\omega) = \partial_t f(t,\omega(t))$ is simply the partial (right) derivative in t: the horizontal derivative is thus an extension of the notion of 'partial derivative in time' for non-anticipative functionals.

5.2.2   Vertical derivative

We now define the *Dupire derivative* or *vertical derivative* of a non-anticipative functional [8, 21]: this derivative captures the sensitivity of a functional to a 'vertical perturbation'

$$\omega_t^e = \omega_t + e1_{[t,T]} \tag{5.4}$$

of a stopped path (t,ω).

Definition 5.2.2 (Vertical derivative [21]). *A non-anticipative functional F is said to be vertically differentiable at $(t,\omega) \in \Lambda_T$ if the map*

$$\mathbb{R}^d \longrightarrow \mathbb{R},$$

$$e \longmapsto F(t,\omega_t + e1_{[t,T]})$$

is differentiable at 0. Its gradient at 0 is called the vertical derivative of F at (t,ω):

$$\nabla_\omega F(t,\omega) = (\partial_i F(t,\omega), \ i = 1, \ldots, d), \quad where$$

$$\partial_i F(t,\omega) = \lim_{h \to 0} \frac{F(t, \omega_t + he_i1_{[t,T]}) - F(t,\omega_t)}{h}. \tag{5.5}$$

If F is vertically differentiable at all $(t,\omega) \in \Lambda_T$, then $\nabla_\omega F$ is a non-anticipative functional called the vertical derivative of F.

For each $e \in \mathbb{R}^d$, $\nabla_\omega F(t,\omega) \cdot e$ is simply the directional derivative of $F(t,\cdot)$ in the direction $1_{[t,T]}e$. A similar notion of functional derivative was introduced by Fliess [30] in the context of optimal control, for causal functionals on bounded variation paths.

Note that $\nabla_\omega F(t,\omega)$ is 'local' in time: $\nabla_\omega F(t,\omega)$ only depends on the partial map $F(t,\cdot)$. However, even if $\omega \in C^0([0,T],\mathbb{R}^d)$, to compute $\nabla_\omega F(t,\omega)$ we need to compute F outside $C^0([0,T],\mathbb{R}^d)$. Also, all terms in (5.5) only depend on ω through $\omega_t = \omega(t \wedge \cdot)$ so, if F is vertically differentiable, then

$$\nabla_\omega F \colon (t,\omega) \longmapsto \nabla_\omega F(t,\omega)$$

defines a non-anticipative functional. This is due to the fact that the perturbations involved only affect the future portion of the path, in contrast, for example, with the Fréchet derivative, which involves perturbating in all directions. One may repeat this operation on $\nabla_\omega F$ and define $\nabla_\omega^2 F, \nabla_\omega^k F, \ldots$. For instance, $\nabla_\omega^2 F(t,\omega)$ is defined as the gradient at 0 (if it exists) of the map

$$e \in \mathbb{R}^d \longmapsto \nabla_\omega F(t, \omega + e\, 1_{[t,T]}).$$

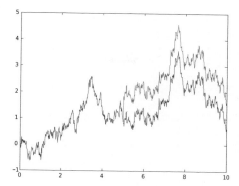

Figure 5.1: The **vertical** perturbation (t, ω_t^e) of the stopped path $(t, \omega) \in \Lambda_T$ in the direction $e \in \mathbb{R}^d$ is obtained by shifting the *future portion of the path* by e: $\omega_t^e = \omega_t + e1_{[t,T]}$.

5.2.3 Regular functionals

A special role is played by non-anticipative functionals which are horizontally differentiable, twice vertically differentiable, and whose derivatives are left-continuous in the sense of Definition 5.1.4 and boundedness preserving (in the sense of Definition 5.1.6):

Definition 5.2.3 ($\mathbb{C}_b^{1,2}$ functionals). *Define* $\mathbb{C}_b^{1,2}(\Lambda_T)$ *as the set of left-continuous functionals* $F \in \mathbb{C}_l^{0,0}(\Lambda_T)$ *such that*

(i) F *admits a horizontal derivative* $\mathcal{D}F(t, \omega)$ *for all* $(t, \omega) \in \Lambda_T$, *and the map* $\mathcal{D}F(t, \cdot) \colon (D([0,T], \mathbb{R}^d, \|.\|_\infty) \to \mathbb{R}$ *is continuous for each* $t \in [0, T[$;

(ii) $\nabla_\omega F, \nabla_\omega^2 F \in \mathbb{C}_l^{0,0}(\Lambda_T)$;

(iii) $\mathcal{D}F, \nabla_\omega F, \nabla_\omega^2 F \in \mathbb{B}(\Lambda_T)$ *(see Definition 5.1.6)*.

Similarly, we can define the class $\mathbb{C}_b^{1,k}(\Lambda_T)$.

Note that this definition only involves *certain* directional derivatives and is therefore much weaker than requiring (even first-order) Fréchet or even Gâteaux differentiability: it is easy to construct examples of $F \in \mathbb{C}_b^{1,2}(\Lambda_T)$ for which even the first order Fréchet derivative does not exist.

Remark 5.2.4. In the proofs of the key results below, one needs either left- or right-continuity of the derivatives, but not both. The pathwise statements of this chapter and the results on functionals of *continuous* semimartingales hold in either case. For functionals of càdlàg semimartingales, however, left- and right-continuity are not interchangeable.

We now give some fundamental examples of classes of smooth functionals.

Example 5.2.5 (Cylindrical non-anticipative functionals). For $g \in C^0(\mathbb{R}^{n \times d})$, $h \in C^k(\mathbb{R}^d)$, with $h(0) = 0$, let

$$F(t, \omega) = h\left(\omega(t) - \omega(t_n-)\right) 1_{t \geq t_n} g(\omega(t_1-), \omega(t_2-), \ldots, \omega(t_n-)).$$

Then, $F \in \mathbb{C}_b^{1,k}(\Lambda_T)$, $\mathcal{D}F(t, \omega) = 0$ and, for every $j = 1, \ldots, k$, we have

$$\nabla_\omega^j F(t, \omega) = \nabla^j h\left(\omega(t) - \omega(t_n-)\right) 1_{t \geq t_n} g\left(\omega(t_1-), \omega(t_2-), \ldots, \omega(t_n-)\right).$$

Example 5.2.6 (Integral functionals). Let $g \in C^0(\mathbb{R}^d)$ and $\rho \colon \mathbb{R}_+ \to \mathbb{R}$ be bounded and measurable. Define

$$F(t, \omega) = \int_0^t g(\omega(u))\rho(u)du. \tag{5.6}$$

Then, $F \in \mathbb{C}_b^{1,\infty}(\Lambda_T)$ with

$$\mathcal{D}F(t, \omega) = g(\omega(t))\rho(t), \quad \nabla_\omega^j F(t, \omega) = 0.$$

Integral functionals of type (5.6) are thus 'purely horizontal' (i.e., they have zero vertical derivative), while cylindrical functionals are 'purely vertical' (they have zero horizontal derivative). In Section 5.3, we will see that any smooth functional may in fact be decomposed into horizontal and vertical components.

Another important class of functionals are conditional expectation operators. We now give an example of smoothness result for such functionals, see [12].

Example 5.2.7 (Weak Euler–Maruyama scheme). Let $\sigma \colon (\Lambda_T, d_\infty) \to \mathbb{R}^{d \times d}$ be a Lipschitz map and W a Wiener process on $(\Omega, \mathcal{F}, \mathbb{P})$. Then the path dependent SDE

$$X(t) = X(0) + \int_0^t \sigma(u, X_u)dW(u) \tag{5.7}$$

has a unique $\mathcal{F}_t^W$-adapted strong solution. The (piecewise constant) Euler–Maruyama approximation for (5.7) then defines a non-anticipative functional $_nX$, given by the recursion

$$_nX(t_{j+1}, \omega) = {}_nX(t_j, \omega) + \sigma\left(t_j, {}_nX_{t_j}(\omega)\right) \cdot \left(\omega(t_{j+1}-) - \omega(t_j-)\right). \tag{5.8}$$

For a Lipschitz functional $g \colon (D([0,T], \mathbb{R}^d), \|.\|_\infty) \to \mathbb{R}$, consider the 'weak Euler approximation'

$$F_n(t, \omega) = \mathbb{E}\left[g\left({}_nX_T(W_T)\right) | \mathcal{F}_t^W\right](\omega) \tag{5.9}$$

for the conditional expectation $\mathbb{E}\left[g(X_T) | \mathcal{F}_t^W\right]$, computed by initializing the scheme on $[0, t]$ with ω, and then iterating (5.8) with the increments of the Wiener process between t and T. Then $F_n \in \mathbb{C}_b^{1,\infty}(\mathcal{W}_T)$, see [12].

This last example implies that a large class of functionals defined as conditional expectations may be approximated in L^p norm by smooth functionals [12]. We will revisit this important point in Chapter 7.

If $F \in \mathbb{C}_b^{1,2}(\Lambda_T)$, then, for any $(t, \omega) \in \Lambda_T$, the map

$$g_{(t,\omega)} : e \in \mathbb{R}^d \longmapsto F(t, \omega + e1_{[t,T]})$$

is twice continuously differentiable on a neighborhood of the origin, and

$$\nabla_\omega F(t, \omega) = \nabla g_{(t,\omega)}(0), \qquad \nabla_\omega^2 F(t, \omega) = \nabla^2 g_{(t,\omega)}(0).$$

A second-order Taylor expansion of the map $g_{(t,\omega)}$ at the origin yields that any $F \in \mathbb{C}_b^{1,2}(\Lambda_T)$ admits a second-order Taylor expansion with respect to a vertical perturbation: $\forall(t, \omega) \in \Lambda_T$, $\forall e \in \mathbb{R}^d$,

$$F(t, \omega + e1_{[t,T]}) = F(t, \omega) + \nabla_\omega F(t, \omega) \cdot e + <e, \nabla_\omega^2 F(t, \omega) \cdot e> + o(\|e\|^2).$$

Schwarz's theorem applied to $g_{(t,\omega)}$ then entails that $\nabla_\omega^2 F(t, \omega)$ is a symmetric matrix. As is well known, the continuity assumption of the derivatives cannot be relaxed in Schwarz's theorem, so one can easily construct counterexamples where $\nabla_\omega^2 F(t, \omega)$ exists but is not symmetric by removing the continuity requirement on $\nabla_\omega^2 F$.

However, unlike the usual partial derivatives in finite dimensions, the horizontal and vertical derivative do not commute, i.e., in general, $\mathcal{D}(\nabla_\omega F) \neq \nabla_\omega(\mathcal{D}F)$.

This stems from the fact that the elementary operations of 'horizontal extension' and 'vertical perturbation' defined above do not commute: a horizontal extension of the stopped path (t, ω) to $t + h$ followed by a vertical perturbation yields the path $\omega_t + e1_{[t+h,T]}$, while a vertical perturbation at t followed by a horizontal extension to $t + h$ yields $\omega_t + e1_{[t,T]} \neq \omega_t + e1_{[t+h,T]}$. Note that these two paths have the same value at $t + h$, so only functionals which are truly path dependent (as opposed to functions of the path at a single point in time) will be affected by this lack of commutativity. Thus, the 'functional Lie bracket'

$$[\mathcal{D}, \nabla_\omega]F = \mathcal{D}(\nabla_\omega F) - \nabla_\omega(\mathcal{D}F)$$

may be used to quantify the 'path dependency' of F.

Example 5.2.8. Let F be the integral functional given in Example 5.2.6, with $g \in C^1(\mathbb{R}^d)$. Then, $\nabla_\omega F(t, \omega) = 0$ so $\mathcal{D}(\nabla_\omega F) = 0$. However, $\mathcal{D}F(t, \omega) = g(\omega(t))$ and hence,

$$\nabla_\omega \mathcal{D}F(t, \omega) = \nabla g(\omega(t)) \neq \mathcal{D}(\nabla_\omega F)(t, \omega) = 0.$$

Locally regular functionals. Many examples of functionals, especially those involving exit times, may fail to be globally smooth, but their derivatives may still be well behaved, except at certain stopping times. The following is a prototypical example of a functional involving exit times.

Example 5.2.9 (A functional involving exit times). Let W be real Brownian motion, $b > 0$, and $M(t) = \sup_{0 \leq s \leq t} W(s)$. Consider the $\mathcal{F}_t^W$-adapted martingale:

$$Y(t) = E[1_{M(T) \geq b} | \mathcal{F}_t^W].$$

Then Y has the functional representation $Y(t) = F(t, W_t)$, where

$$F(t, \omega) = 1_{\sup_{0 \leq s \leq t} \omega(s) \geq b} + 1_{\sup_{0 \leq s \leq t} \omega(s) < b} \left[2 - 2N \left(\frac{b - \omega(t)}{\sqrt{T - t}} \right) \right],$$

and where N is the $N(0, 1)$ distribution. Observe that $F \notin \mathbb{C}_l^{0,0}(\Lambda_T)$: a path ω_t where $\omega(t) < b$, but $\sup_{0 \leq s \leq t} \omega(s) = b$ can be approximated in sup norm by paths where $\sup_{0 \leq s \leq t} \omega(s) < b$. However, one can easily check that $\nabla_\omega F$, $\nabla_\omega^2 F$, and $\mathcal{D}F$ exist almost everywhere.

We recall that a stopping time (or a non-anticipative random time) on $(\Omega, (\mathcal{F}_t^0)_{t \in [0,T]})$ is a measurable map $\tau : \Omega \to [0, \infty)$ such that $\forall t \geq 0$,

$$\{\omega \in \Omega, \tau(\omega) \leq t\} \in \mathcal{F}_t^0.$$

In the example above, regularity holds except on the graph of a certain stopping time. This motivates the following definition:

Definition 5.2.10 ($\mathbb{C}_{\mathrm{loc}}^{1,2}(\Lambda_T)$). $F \in \mathbb{C}_b^{0,0}(\Lambda_T)$ *is said to be* locally regular *if there exists an increasing sequence* $(\tau_k)_{k \geq 0}$ *of stopping times with* $\tau_0 = 0$, $\tau_k \uparrow \infty$, *and* $F^k \in \mathbb{C}_b^{1,2}(\Lambda_T)$ *such that*

$$F(t, \omega) = \sum_{k \geq 0} F^k(t, \omega) 1_{[\tau_k(\omega), \tau_{k+1}(\omega)[}(t).$$

Note that $\mathbb{C}_b^{1,2}(\Lambda_T) \subset \mathbb{C}_{\mathrm{loc}}^{1,2}(\Lambda_T)$, but the notion of local regularity allows discontinuities or explosions at the times described by $(\tau_k, k \geq 1)$.

Revisiting Example 5.2.9, we can see that Definition 5.2.10 applies: recall that

$$F(t, \omega) = 1_{\sup_{0 \leq s \leq t} \omega(s) \geq b} + 1_{\sup_{0 \leq s \leq t} \omega(s) < b} \left[2 - 2N \left(\frac{b - \omega(t)}{\sqrt{T - t}} \right) \right].$$

If we define

$$\tau_1(\omega) = \inf\{t \geq 0 | \omega(t) = b\} \wedge T, \quad F^0(t, \omega) = 2 - 2N \left(\frac{b - \omega(t)}{\sqrt{T - t}} \right),$$

$$\tau_i(\omega) = T + i - 2, \quad \text{for} \quad i \geq 2; \quad F^i(t, \omega) = 1, \quad i \geq 1,$$

then $F^i \in \mathbb{C}_b^{1,2}(\Lambda_T)$, so $F \in \mathbb{C}_{\mathrm{loc}}^{1,2}(\Lambda_T)$.

5.3 Pathwise integration and functional change of variable formula

In the seminal paper [31], Föllmer proposed a non-probabilistic version of the Itô formula: Föllmer showed that if a càdlàg (right continuous with left limits) function x has finite quadratic variation along some sequence $\pi_n = (0 = t_0^n < t_1^n < \cdots < t_n^n = T)$ of partitions of $[0, T]$ with step size decreasing to zero, then for $f \in C^2(\mathbb{R}^d)$ one can define the pathwise integral

$$\int_0^T \nabla f(x(t)) d^\pi x = \lim_{n \to \infty} \sum_{i=0}^{n-1} \nabla f\big(x(t_i^n)\big) \cdot \big(x(t_{i+1}^n) - x(t_i^n)\big) \qquad (5.10)$$

as a limit of Riemann sums along the sequence $\pi = (\pi_n)_{n \geq 1}$ of partitions, and obtain a change of variable formula for this integral, see [31]. We now revisit Föllmer's approach and show how may it be combined with Dupire's directional derivatives to obtain a pathwise change of variable formula for functionals in $\mathbb{C}_{\mathrm{loc}}^{1,2}(\Lambda_T)$.

5.3.1 Quadratic variation of a path along a sequence of partitions

We first define the notion of quadratic variation of a path along a sequence of partitions. Our presentation is different from Föllmer [31], but can be shown to be mathematically equivalent.

Throughout this subsection $\pi = (\pi_n)_{n \geq 1}$ will denote a sequence of partitions of $[0, T]$ into intervals:

$$\pi_n = \big(0 = t_0^n < t_1^n < \cdots < t_{k(n)}^n = T\big),$$

and $|\pi_n| = \sup\{|t_{i+1}^n - t_i^n|, \ i = 1, \ldots, k(n)\}$ will denote the mesh size of the partition π_n.

As an example, one can keep in mind the *dyadic partition*, for which $t_i^n = iT/2^n$, $i = 0, \ldots, k(n) = 2^n$, and $|\pi_n| = 2^{-n}$. This is an example of a *nested* sequence of partitions: for $n \geq m$, every interval $[t_i^n, t_{i+1}^n]$ of the partition π_n is included in one of the intervals of π_m. Unless specified, we will assume that the sequence π_n is a nested sequence of partitions.

Definition 5.3.1 (Quadratic variation of a path along a sequence of partitions). *Let* $\pi_n = (0 = t_0^n < t_1^n < \cdots < t_{k(n)}^n = T)$ *be a sequence of partitions of* $[0, T]$ *with step size decreasing to zero. A path* $x \in D([0, T], \mathbb{R})$ *is said to have* finite quadratic *variation along the sequence of partitions* $(\pi_n)_{n \geq 1}$ *if for any* $t \in [0, T]$ *the limit*

$$[x]_\pi(t) := \lim_{n \to \infty} \sum_{t_{i+1}^n \leq t} \big(x(t_{i+1}^n) - x(t_i^n)\big)^2 < \infty \qquad (5.11)$$

exists and the increasing function $[x]$ *has a Lebesgue decomposition*

$$[x]_\pi(t) = [x]_\pi^c(t) + \sum_{0 < s \leq t} |\Delta x(s)|^2,$$

where $[x]_\pi^c$ is a continuous, increasing function.

The increasing function $[x]_\pi \colon [0,T] \to \mathbb{R}_+$ defined by (5.11) is then called the quadratic variation of the path x along the sequence of partitions $\pi = (\pi_n)_{n \geq 1}$, and $[x]_\pi^c$ *is the continuous quadratic variation of x along π. We denote by $Q^\pi([0,T],\mathbb{R})$ the set of càdlàg paths with finite quadratic variation along the sequence of partitions $\pi = (\pi_n)_{n \geq 1}$.*

In general, the quadratic variation of a path x along a sequence of partitions π depends on the choice of the sequence π, as the following example shows.

Example 5.3.2. Let $\omega \in C^0([0,T],\mathbb{R})$ be an arbitrary continuous function. Let us construct recursively a sequence π_n of partitions of $[0,T]$ such that

$$|\pi_n| \leq \frac{1}{n} \quad \text{and} \quad \sum_{\pi_n} |\omega(t_{k+1}^n) - \omega(t_k^n)|^2 \leq \frac{1}{n}.$$

Assume we have constructed π_n with the above property. Adding to π_n the points $kT/(n+1)$, $k = 1, \ldots, n$, we obtain a partition $\sigma_n = (s_i^n, i = 0, \ldots, M_n)$ with $|\sigma_n| \leq 1/(n+1)$. For $i = 0, \ldots, (M_n - 1)$, we further refine $[s_i^n, s_{i+1}^n]$ as follows. Let $J(i)$ be an integer with

$$J(i) \geq (n+1)M_n |\omega(s_{i+1}^n) - \omega(s_i^n)|^2,$$

and define $\tau_{i,1}^n = s_i^n$ and, for $k = 1, \ldots, J(i)$,

$$\tau_{i,k+1}^n = \inf\left\{ t \geq \tau_{i,k}^n, \quad \omega(t) = \omega(s_i^n) + \frac{k\left(\omega(s_{i+1}^n) - \omega(s_i^n)\right)}{J(i)} \right\}.$$

Then the points $(\tau_{i,k}^n, k = 1, \ldots, J(i))$, define a partition of $[s_i^n, s_{i+1}^n]$ with

$$|\tau_{i,k+1}^n - \tau_{i,k}^n| \leq \frac{1}{n+1} \quad \text{and} \quad |\omega(\tau_{i,k+1}^n) - \omega(\tau_{i,k}^n)| = \frac{|\omega(s_{i+1}^n) - \omega(s_i^n)|}{J(i)},$$

so

$$\sum_{k=1}^{J(i)} |\omega(\tau_{i,k+1}^n) - \omega(\tau_{i,k}^n)|^2 \leq J(i)\frac{|\omega(s_{i+1}^n) - \omega(s_i^n)|^2}{J(i)^2} = \frac{1}{(n+1)M_n}.$$

Sorting $(\tau_{i,k}^n, i = 0, \ldots, M_n, k = 1, \ldots, J(i))$ in increasing order, we thus obtain a partition $\pi_{n+1} = (t_j^{n+1})$ of $[0,T]$ such that

$$|\pi_{n+1}| \leq \frac{1}{n+1}, \quad \sum_{\pi_{n+1}} |\omega(t_{i+1}^n) - \omega(t_i^n)|^2 \leq \frac{1}{n+1}.$$

Taking limits along this sequence $\pi = (\pi_n, n \geq 1)$ then yields $[\omega]_\pi = 0$.

This example shows that "having finite quadratic variation along *some* sequence of partitions" is not an interesting property, and that the notion of quadratic variation along a sequence of partitions depends on the chosen partition. Definition 5.3.1 becomes non-trivial only if one *fixes* the partition beforehand. In the sequel, we fix a sequence $\pi = (\pi_n, n \geq 1)$ of partitions with $|\pi_n| \to 0$ and all limits will be considered along the *same* sequence π, thus enabling us to drop the subscript in $[x]_\pi$ whenever the context is clear.

The notion of quadratic variation along a sequence of partitions is different from the *p*-variation of the path ω for $p = 2$: the *p*-variation involves taking a supremum over *all* partitions, not necessarily formed of intervals, whereas (5.11) is a limit taken along a given sequence $(\pi_n)_{n \geq 1}$. In general $[x]_\pi$, given by (5.11), is smaller than the 2-variation and there are many situations where the 2-variation is infinite, while the limit in (5.11) is finite. This is in fact the case, for instance, for typical paths of Brownian motion, which have finite quadratic variation along any sequence of partitions with mesh size $o(1/\log n)$ [20], but have infinite *p*-variation almost surely for $p \leq 2$, see [72].

The extension of this notion to vector-valued paths is somewhat subtle, see [31].

Definition 5.3.3. *A d-dimensional path* $x = (x^1, \ldots, x^d) \in D([0, T], \mathbb{R}^d)$ *is said to have* finite quadratic variation along $\pi = (\pi_n)_{n \geq 1}$ *if* $x^i \in Q^\pi([0, T], \mathbb{R})$ *and* $x^i + x^j \in Q^\pi([0, T], \mathbb{R})$ *for all* $i, j = 1, \ldots, d$. *Then, for any* $i, j = 1, \ldots, d$ *and* $t \in [0, T]$, *we have*

$$\sum_{t_k^n \in \pi_n, t_k^n \leq t} \left(x^i(t_{k+1}^n) - x^i(t_k^n) \right) \cdot \left(x^j(t_{k+1}^n) - x^j(t_k^n) \right) \xrightarrow{n \to \infty} [x]_{ij}(t)$$

$$= \frac{[x^i + x^j](t) - [x^i](t) - [x^j](t)}{2}.$$

The matrix-valued function $[x]: [0, T] \to S_d^+$ *whose elements are given by*

$$[x]_{ij}(t) = \frac{[x^i + x^j](t) - [x^i](t) - [x^j](t)}{2}$$

is called the quadratic covariation *of the path* x*: for any* $t \in [0, T]$,

$$\sum_{t_i^n \in \pi_n, t_n \leq t} \left(x(t_{i+1}^n) - x(t_i^n) \right) \cdot^t \left(x(t_{i+1}^n) - x(t_i^n) \right) \xrightarrow{n \to \infty} [x](t) \in S_d^+,$$

and $[x]$ *is increasing in the sense of the order on positive symmetric matrices: for* $h \geq 0$, $[x](t + h) - [x](t) \in S_d^+$.

We denote by $Q^\pi([0, T], \mathbb{R}^d)$ the set of $\mathbb{R}^d$-valued càdlàg paths with finite quadratic variation with respect to the sequence of partitions $\pi = (\pi_n)_{n \geq 1}$.

Remark 5.3.4. Note that Definition 5.3.3 requires that $x^i + x^j \in Q^\pi([0,T],\mathbb{R})$; this does not necessarily follow from requiring $x^i, x^j \in Q^\pi([0,T],\mathbb{R})$. Indeed, denoting $\delta x^i = x^i(t_{k+1}) - x^i(t_k)$, we have

$$\left|\delta x^i + \delta x^j\right|^2 = |\delta x^i|^2 + |\delta x^j|^2 + 2\delta x^i \delta x^j,$$

and the cross-product may be positive, negative, or have an oscillating sign which may prevent convergence of the series $\sum_{\pi_n} \delta x^i \delta x^j$ [64]. However, if x^i, x^j are differentiable functions of the *same* path ω, i.e., $x^i = f_i(\omega)$ with $f_i \in C^1(\mathbb{R}^d, \mathbb{R})$, then

$$\delta x^i \delta x^j = f_i'(\omega(t_k^n)) f_j'(\omega(t_k^n)) |\delta \omega|^2 + o(|\delta \omega|^2)$$

so, $\sum_{\pi_n} \delta x^i \delta x^j$ converges and

$$\lim_{n \to \infty} \sum_{\pi_n} \delta x^i \delta x^j = [x^i, x^j]$$

is well-defined. This remark is connected to the notion of 'controlled rough path' introduced by Gubinelli [35], see Remark 5.3.9 below.

For any $x \in Q^\pi([0,T],\mathbb{R}^d)$, since $[x]\colon [0,T] \to S_d^+$ is increasing (in the sense of the order on S_d^+), we can define the Riemann–Stieltjes integral $\int_0^T f d[x]$ for any $f \in C_b^0([0,T])$. A key property of $Q^\pi([0,T],\mathbb{R}^d)$ is the following.

Proposition 5.3.5 (Uniform convergence of quadratic Riemann sums). *For every* $\omega \in Q^\pi([0,T], \mathbb{R}^d)$, *every* $h \in C_b^0([0,T], \mathbb{R}^{d \times d})$, *and every* $t \in [0,T]$, *we have*

$$\sum_{t_i^n \in \pi_n, t_i^n \le t} \mathrm{tr}\big(h(t_i^n)(\omega(t_{i+1}^n) - \omega(t_i^n))^t (\omega(t_{i+1}^n) - \omega(t_i^n))\big) \overset{n \to \infty}{\longrightarrow} \int_0^t \langle h, d[\omega] \rangle,$$

where we use the notation $\langle A, B \rangle = \mathrm{tr}(^t A \cdot B)$ *for* $A, B \in \mathbb{R}^{d \times d}$.
Furthermore, if $\omega \in C^0([0,T], \mathbb{R}^d)$, *the convergence is uniform in* $t \in [0,T]$.

Proof. It suffices to show this property for $d = 1$. Let $\omega \in Q^\pi([0,T],\mathbb{R})$ and $h \in D([0,T],\mathbb{R})$. Then the integral $\int_0^t h \, d[\omega]$ may be defined as a limit of Riemann sums:

$$\int_0^t h d[\omega] = \lim_{n \to \infty} \sum_{\pi_n} h(t_i^n)([\omega](t_{i+1}^n) - [\omega](t_i^n)).$$

Using the definition of $[\omega]$, the result is thus true for $h\colon [0,T] \to \mathbb{R}$ of the form $h = \sum_{\pi_n} a_k 1_{[t_k^n, t_{k+1}^n[}$. Consider now $h \in C_b^0([0,T],\mathbb{R})$ and define the piecewise constant approximations

$$h^n = \sum_{\pi_n} h(t_k^n) 1_{[t_k^n, t_{k+1}^n[}.$$

Then h^n converges uniformly to h: $\|h - h^n\|_\infty \to 0$ as $n \to \infty$ and, for each $n \geq 1$,

$$\sum_{t_i^k \in \pi_k, t_i^k \leq t} h^n(t_i^k)\big(\omega(t_{i+1}^k) - \omega(t_i^k)\big)^2 \overset{k \to \infty}{\longrightarrow} \int_0^t h^n d[\omega].$$

Since h is bounded on $[0, T]$, this sequence is dominated. We can then conclude using a diagonal convergence argument. $\square$

Proposition 5.3.5 implies weak convergence on $[0, T]$ of the discrete measures

$$\xi^n = \sum_{i=0}^{k(n)-1} \big(\omega(t_{i+1}^n) - \omega(t_i^n)\big)^2 \delta_{t_i^n} \overset{n \to \infty}{\Longrightarrow} \xi = d[\omega],$$

where δ_t is the Dirac measure (point mass) at t.[1]

5.3.2 Functional change of variable formula

Consider now $\omega \in Q^\pi([0, T], \mathbb{R}^d)$. Since ω has at most a countable set of jump times, we may always assume that the partition 'exhausts' the jump times in the sense that

$$\sup_{t \in [0,T] - \pi_n} |\omega(t) - \omega(t-)| \overset{n \to \infty}{\longrightarrow} 0. \tag{5.12}$$

Then the piecewise constant approximation

$$\omega^n(t) = \sum_{i=0}^{k(n)-1} \omega(t_{i+1}-)1_{[t_i, t_{i+1}[}(t) + \omega(T)1_{\{T\}}(t) \tag{5.13}$$

converges uniformly to ω:

$$\sup_{t \in [0,T]} \|\omega^n(t) - \omega(t)\| \overset{n \to \infty}{\longrightarrow} 0.$$

By decomposing the variations of the functional into vertical and horizontal increments along the partition π_n, we obtain the following result:

Theorem 5.3.6 (Pathwise change of variable formula for $\mathbb{C}^{1,2}$ functionals, [8]). *Let* $\omega \in Q^\pi([0, T], \mathbb{R}^d)$ *verifying* (5.12). *Then, for any* $F \in \mathbb{C}^{1,2}_{\text{loc}}(\Lambda_T)$, *the limit*

$$\int_0^T \nabla_\omega F(t, \omega_{t-}) d^\pi \omega := \lim_{n \to \infty} \sum_{i=0}^{k(n)-1} \nabla_\omega F(t_i^n, \omega_{t_i^n}^n)\big(\omega(t_{i+1}^n) - \omega(t_i^n)\big) \tag{5.14}$$

---

[1]In fact, this weak convergence property was used by Föllmer [31] as the definition of pathwise quadratic variation. We use the more natural definition (Definition 5.3.1) of which Proposition 5.3.5 is a consequence.

exists, and

$$F(T, \omega_T) - F(0, \omega_0) = \int_0^T \mathcal{D}F(t, \omega_t) dt + \int_0^T \frac{1}{2} \text{tr} \left({}^t \nabla_\omega^2 F(t, \omega_{t-}) d[\omega]^c(t) \right)$$

$$+ \int_0^T \nabla_\omega F(t, \omega_{t-}) d^\pi \omega + \sum_{t \in]0,T]} [F(t, \omega_t) - F(t, \omega_{t-}) - \nabla_\omega F(t, \omega_{t-}) \cdot \Delta\omega(t)].$$

The detailed proof of Theorem 5.3.6 may be found in [8] under more general assumptions. Here, we reproduce a simplified version of this proof in the case where ω is continuous.

Proof of Theorem 5.3.6. First we note that, up to localization by a sequence of stopping times, we can assume that $F \in \mathbb{C}_b^{1,2}(\Lambda_T)$, which we shall do in the sequel. Denote $\delta\omega_i^n = \omega(t_{i+1}^n) - \omega(t_i^n)$. Since ω is continuous on $[0, T]$, it is uniformly continuous, so

$$\eta_n = \sup \left\{ |\omega(u) - \omega(t_{i+1}^n)| + |t_{i+1}^n - t_i^n|, \ 0 \le i \le k(n) - 1, \ u \in [t_i^n, t_{i+1}^n] \right\} \overset{n\to\infty}{\longrightarrow} 0.$$

Since $\nabla_\omega^2 F, \mathcal{D}F$ satisfy the boundedness preserving property (5.2), for n sufficiently large there exists $C > 0$ such that for every $t < T$ and for every $\omega' \in \Lambda_T$,

$$d_\infty \left((t, \omega), (t', \omega') \right) < \eta_n \implies |\mathcal{D}F(t', \omega')| \le C, \ |\nabla_\omega^2 F(t', \omega')| \le C.$$

For $i \le k(n) - 1$, consider the decomposition of increments into 'horizontal' and 'vertical' terms:

$$F\left(t_{i+1}^n, \omega_{t_{i+1}^n}^n -\right) - F\left(t_i^n, \omega_{t_i^n}^n -\right) = F\left(t_{i+1}^n, \omega_{t_{i+1}^n}^n -\right) - F\left(t_i^n, \omega_{t_i^n}^n\right)$$

$$+ F(t_i^n, \omega_{t_i^n}^n) - F(t_i^n, \omega_{t_i^n}^n -). \tag{5.15}$$

The first term in (5.15) can be written $\psi(h_i^n) - \psi(0)$, where $h_i^n = t_{i+1}^n - t_i^n$ and

$$\psi(u) = F\left(t_i^n + u, \omega_{t_i^n}^n\right).$$

Since $F \in \mathbb{C}_b^{1,2}(\Lambda_T)$, ψ is right differentiable. Moreover, by Proposition 5.1.5, ψ is left continuous so,

$$F\left(t_{i+1}^n, \omega_{t_i^n}^n\right) - F\left(t_i^n, \omega_{t_i^n}^n\right) = \int_0^{t_{i+1}^n - t_i^n} \mathcal{D}F\left(t_i^n + u, \omega_{t_i^n}^n\right) du.$$

The second term in (5.15) can be written $\phi(\delta\omega_i^n) - \phi(0)$, where

$$\phi(u) = F\left(t_i^n, \omega_{t_i^n}^n - + u 1_{[t_i^n, T]}\right).$$

Since $F \in \mathbb{C}_b^{1,2}(\Lambda_T)$, $\phi \in C^2(\mathbb{R}^d)$ with

$$\nabla\phi(u) = \nabla_\omega F\left(t_i^n, \omega_{t_i^n}^n - + u 1_{[t_i^n, T]}\right), \quad \nabla^2\phi(u) = \nabla_\omega^2 F\left(t_i^n, \omega_{t_i^n}^n - + u 1_{[t_i^n, T]}\right).$$

A second-order Taylor expansion of ϕ at $u = 0$ yields

$$F\left(t_i^n, \omega_{t_i^n}^n\right) - F\left(t_i^n, \omega_{t_i^n-}^n\right) = \nabla_\omega F\left(t_i^n, \omega_{t_i^n-}^n\right) \delta\omega_i^n$$

$$+ \frac{1}{2} \mathrm{tr}\left(\nabla_\omega^2 F\left(t_i^n, \omega_{t_i^n-}^n\right) \cdot^t \delta\omega_i^n \delta\omega_i^n\right) + r_i^n,$$

where r_i^n is bounded by

$$K\left|\delta\omega_i^n\right|^2 \sup_{x \in B(0,\eta_n)} \left|\nabla_\omega^2 F\left(t_i^n, \omega_{t_i^n-}^n + x 1_{[t_i^n, T]}\right) - \nabla_\omega^2 F\left(t_i^n, \omega_{t_i^n-}^n\right)\right|.$$

Denote $i^n(t)$ the index such that $t \in [t_{i^n(t)}^n, t_{i^n(t)+1}^n[$. We now sum all the terms above from $i = 0$ to $k(n) - 1$:

- The left-hand side of (5.15) yields $F(T, \omega_{T-}^n) - F(0, \omega_0^n)$, which converges to $F(T, \omega_{T-}) - F(0, \omega_0)$ by left continuity of F, and this quantity equals $F(T, \omega_T) - F(0, \omega_0)$ since ω is continuous.

- The first line in the right-hand side can be written

$$\int_0^T \mathcal{D}F\left(u, \omega_{t_{i^n(u)}^n}^n\right) du, \tag{5.16}$$

where the integrand converges to $\mathcal{D}F(u, \omega_u)$ and is bounded by C. Hence, the dominated convergence theorem applies and (5.16) converges to

$$\int_0^T \mathcal{D}F(u, \omega_u) du$$

- The second line can be written

$$\sum_{i=0}^{k(n)-1} \nabla_\omega F\left(t_i^n, \omega_{t_i^n-}^n\right) \cdot \left(\omega(t_{i+1}^n) - \omega(t_i^n)\right)$$

$$+ \sum_{i=0}^{k(n)-1} \frac{1}{2} \mathrm{tr}\left(\nabla_\omega^2 F(t_i^n (\omega_{t_i^n-}^n)^t \delta\omega_i^n \delta\omega_i^n\right) + \sum_{i=0}^{k(n)-1} r_i^n.$$

The term $\nabla_\omega^2 F(t_i^n, \omega_{t_i^n-}^n) 1_{]t_i^n, t_{i+1}^n]}$ is bounded by C and, by left continuity of $\nabla_\omega^2 F$, converges to $\nabla_\omega^2 F(t, \omega_t)$; the paths of both are left continuous by Proposition 5.1.5. We now use a 'diagonal lemma' for weak convergence of measures [8].

Lemma 5.3.7 (Cont–Fournié [8]). *Let $(\mu_n)_{n \geq 1}$ be a sequence of Radon measures on $[0, T]$ converging weakly to a Radon measure μ with no atoms, and let $(f_n)_{n \geq 1}$ and f be left-continuous functions on $[0, T]$ such that, for every $t \in [0, T]$,*

$$\lim_n f_n(t) = f(t) \quad \text{and} \quad \|f_n(t)\| \leq K.$$

Then,

$$\int_s^t f_n d\mu_n \overset{n\to\infty}{\Longrightarrow} \int_s^t f d\mu.$$

Applying the above lemma to the second term in the sum, we obtain:

$$\int_0^T \frac{1}{2}\mathrm{tr}\big({}^t\nabla_\omega^2 F(t_i^n, \omega_{t_i^n}^n-)d\xi^n\big) \overset{n\to\infty}{\Longrightarrow} \int_0^T \frac{1}{2}\mathrm{tr}\big({}^t\nabla_\omega^2 F(u,\omega_u)\, d[\omega](u)\big).$$

Using the same lemma, since $|r_i^n|$ is bounded by $\epsilon_i^n|\delta\omega_i^n|^2$, where ϵ_i^n converges to 0 and is bounded by C,

$$\sum_{i=i^n(s)+1}^{i^n(t)-1} r_i^n \overset{n\to\infty}{\longrightarrow} 0.$$

Since all other terms converge, the limit

$$\lim_{n\to\infty} \sum_{i=0}^{k(n)-1} \nabla_\omega F\big(t_i^n, \omega_{t_i^n}^n-\big)\big(\omega(t_{i+1}^n) - \omega(t_i^n)\big)$$

exists.   □

5.3.3 Pathwise integration for paths of finite quadratic variation

A byproduct of Theorem 5.3.6 is that we can define, for $\omega \in Q^\pi([0,T],\mathbb{R}^d)$, the pathwise integral $\int_0^T \phi d^\pi\omega$ as a limit of non-anticipative Riemann sums

$$\int_0^T \phi\, d^\pi\omega := \lim_{n\to\infty} \sum_{i=0}^{k(n)-1} \phi\big(t_i^n, \omega_{t_i^n}^n\big)\big(\omega(t_{i+1}^n) - \omega(t_i^n)\big),$$

for any integrand of the form

$$\phi(t,\omega) = \nabla_\omega F(t,\omega),$$

where $F \in \mathbb{C}_{\mathrm{loc}}^{1,2}(\Lambda_T)$, without requiring that ω be of finite variation. This construction extends Föllmer's pathwise integral, defined in [31] for integrands of the form $\phi = \nabla f \circ \omega$ with $f \in C^2(\mathbb{R}^d)$, to path dependent integrands of the form $\phi(t,\omega)$, where ϕ belongs to the space

$$V(\Lambda_T) = \{\nabla_\omega F(\cdot,\cdot),\ F \in \mathbb{C}_b^{1,2}(\Lambda_T^d)\}. \tag{5.17}$$

Here, Λ_T^d denotes the space of $\mathbb{R}^d$-valued stopped càdlàg paths. We call such integrands *vertical 1-forms*. Since, as noted before, the horizontal and vertical derivatives do not commute, $V(\Lambda_T^d)$ does not coincide with $\mathbb{C}_b^{1,1}(\Lambda_T^d)$.

This set of integrands has a natural vector space structure and includes as subsets the space $\mathbb{S}(\pi, \Lambda_T^d)$ of simple predictable cylindrical functionals as well as Föllmer's space of integrands $\{\nabla f, f \in C^2(\mathbb{R}^d, \mathbb{R})\}$. For $\phi = \nabla_\omega F \in V(\Lambda_T^d)$, the pathwise integral $\int_0^t \phi(t, \omega_{t-}) d^\pi \omega$ is, in fact, given by

$$\int_0^T \phi(t, \omega_{t-}) d^\pi \omega = F(T, \omega_T) - F(0, \omega_0) - \frac{1}{2} \int_0^T \langle \nabla_\omega \phi(t, \omega_{t-}), d[\omega]_\pi \rangle$$

$$- \int_0^T \mathcal{D}F(t, \omega_{t-}) dt - \sum_{0 \leq s \leq T} F(t, \omega_t) - F(t, \omega_{t-}) - \phi(t, \omega_{t-}) \cdot \Delta\omega(t).$$

$$(5.18)$$

The following proposition, whose proof is given in [6], summarizes some key properties of this integral.

Proposition 5.3.8. *Let $\omega \in Q^\pi([0, T], \mathbb{R}^d)$. The pathwise integral (5.18) defines a map*

$$I_\omega : V(\Lambda_T^d) \longrightarrow Q^\pi([0, T], \mathbb{R}^d),$$

$$\phi \longmapsto \int_0^\cdot \phi(t, \omega_{t-}) d^\pi \omega(t)$$

with the following properties:

(i) *pathwise isometry formula: $\forall \phi \in \mathbb{S}(\pi, \Lambda_T^d)$, $\forall s \in [0, T]$,*

$$[I_\omega(\phi)]_\pi(s) = \left[\int_0^\cdot \phi(t, \omega_{t-}) d^\pi \omega \right]_\pi(s) = \int_0^t \langle \phi(t, \omega_{t-})^t \phi(t, \omega_{t-}), d[\omega] \rangle;$$

(ii) *quadratic covariation formula: for $\phi, \psi \in \mathbb{S}(\pi, \Lambda_T^d)$, the limit*

$$[I_\omega(\phi), I_\omega(\psi)] := \lim_{n \to \infty} \sum_{\pi_n} \left(I_\omega(\phi)(t_{k+1}^n) - I_\omega(\phi(t_k^n)) \right) \cdot \left(I_\omega(\psi)(t_{k+1}^n) - I_\omega(\psi)(t_k^n) \right)$$

exists and is given by

$$[I_\omega(\phi), I_\omega(\psi)] = \int_0^\cdot \langle \psi(t, \omega_{t-})^t \phi(t, \omega_{t-}), d[\omega] \rangle;$$

(iii) *associativity: let $\phi \in V(\Lambda_T^d)$, $\psi \in V(\Lambda_T^1)$, and $x \in D([0, T], \mathbb{R})$ defined by $x(t) = \int_0^t \phi(t, \omega_{t-}) d^\pi \omega$. Then,*

$$\int_0^t \psi(t, x_{t-}) d^\pi x = \int_0^t \psi(t, (\int_0^t \phi(u, \omega_{u-}) d^\pi \omega)_{t-}) \phi(t, \omega_{t-}) d^\pi \omega.$$

This pathwise integration has interesting applications in Mathematical Finance [13, 32] and stochastic control [32], where integrals of such vertical 1-forms naturally appear as hedging strategies [13], or optimal control policies and pathwise interpretations of quantities are arguably necessary to interpret the results in terms of the original problem.

Thus, unlike $Q^\pi([0, T], \mathbb{R}^d)$ itself, which does not have a vector space structure, the image $\mathbb{C}_b^{1,2}(\omega)$ of ω by regular functionals is a vector space of paths with finite quadratic variation whose properties are 'controlled' by ω, on which the quadratic variation along the sequence of partitions π is well defined. As argued in [6], this space, and not $Q^\pi([0, T], \mathbb{R}^d)$, is the appropriate setting for studying the pathwise calculus developed here.

Remark 5.3.9 (Relation with 'rough path' theory). For $F \in \mathbb{C}_b^{1,2}$, integrands of the form $\nabla_\omega F(t, \omega)$ with may be viewed as 'controlled rough paths' in the sense of Gubinelli [35]: their increments are 'controlled' by those of ω. However, unlike the approach of rough path theory [35, 49], the pathwise integration defined here does not resort to the use of p-variation norms on iterated integrals: convergence of Riemann sums is pointwise (and, for continuous paths, uniform in t). The reason is that the obstruction to integration posed by the Lévy area, which is the focus of rough path theory, vanishes when considering integrands which are vertical 1-forms. Fortunately, all integrands one encounters when applying the change of variable formula, as well as in applications involving optimal control, hedging, etc, are precisely of the form (5.17). This observation simplifies the approach and, importantly, yields an integral which may be expressed as a limit of (ordinary) Riemann sums, which have an intuitive (physical, financial, etc.) interpretation, without resorting to 'rough path' integrals, whose interpretation is less intuitive.

If ω has finite variation, then the Föllmer integral reduces to the Riemann–Stieltjes integral and we obtain the following result.

Proposition 5.3.10. *For any $F \in \mathbb{C}_{\mathrm{loc}}^{1,1}(\Lambda_T)$ and any $\omega \in BV([0, T]) \cap D([0, T], \mathbb{R}^d)$,*

$$F(T, \omega_T) - F(0, \omega_0) = \int_0^T \mathcal{D}F(t, \omega_{t-})du + \int_0^T \nabla_\omega F(t, \omega_{t-})d\omega$$

$$+ \sum_{t \in]0, T]} [F(t, \omega_t) - F(t, \omega_{t-}) - \nabla_\omega F(t, \omega_{t-}) \cdot \Delta\omega(t)],$$

where the integrals are defined as limits of Riemann sums along any sequence of partitions $(\pi_n)_{n \geq 1}$ with $|\pi_n| \to 0$.

In particular, if ω is continuous with finite variation, we have that, for every $F \in \mathbb{C}_{\mathrm{loc}}^{1,1}(\Lambda_T)$ and every $\omega \in BV([0, T]) \cap C^0([0, T], \mathbb{R}^d)$,

$$F(T, \omega_T) - F(0, \omega_0) = \int_0^T \mathcal{D}F(t, \omega_t)dt + \int_0^T \nabla_\omega F(t, \omega_t)d\omega.$$

Thus, the restriction of any functional $F \in \mathbb{C}^{1,1}_{\mathrm{loc}}(\Lambda_T)$ to $BV([0,T]) \cap C^0([0,T], \mathbb{R}^d)$ may be decomposed into 'horizontal' and 'vertical' components.

5.4 Functionals defined on continuous paths

Consider now an $\mathbb{F}$-adapted process $(Y(t))_{t \in [0,T]}$ given by a functional representation

$$Y(t) = F(t, X_t), \qquad (5.19)$$

where $F \in \mathbb{C}^{0,0}_l(\Lambda_T)$ has left-continuous horizontal and vertical derivatives $\mathcal{D}F \in \mathbb{C}^{0,0}_l(\Lambda_T)$ and $\nabla_\omega F \in \mathbb{C}^{0,0}_l(\Lambda_T)$.

If the process X has continuous paths, Y only depends on the restriction of F to

$$\mathcal{W}_T = \{(t, \omega) \in \Lambda_T, \omega \in C^0([0,T], \mathbb{R}^d)\},$$

so the representation (5.19) is not unique. However, the definition of $\nabla_\omega F$ (Definition 5.2.2), which involves evaluating F on paths to which a jump perturbation has been added, seems to depend on the values taken by F *outside* $\mathcal{W}_T$. It is crucial to resolve this point if one is to deal with functionals of continuous processes or, more generally, processes for which the topological support of the law is not the full space $D([0,T], \mathbb{R}^d)$; otherwise, the very definition of the vertical derivative becomes ambiguous.

This question is resolved by Theorems 5.4.1 and 5.4.2 below, both derived in [10].

The first result below shows that if $F \in \mathbb{C}^{1,1}_l(\Lambda_T)$, then $\nabla_\omega F(t, X_t)$ is uniquely determined by the restriction of F to continuous paths.

Theorem 5.4.1 (Cont and Fournié [10]). *Consider $F^1, F^2 \in \mathbb{C}^{1,1}_l(\Lambda_T)$ with left-continuous horizontal and vertical derivatives. If F^1 and F^2 coincide on continuous paths, i.e.,*

$$\forall t \in [0,T[, \quad \forall \omega \in C^0([0,T], \mathbb{R}^d), \quad F^1(t, \omega_t) = F^2(t, \omega_t),$$

then

$$\forall t \in [0,T[, \quad \forall \omega \in C^0([0,T], \mathbb{R}^d), \quad \nabla_\omega F^1(t, \omega_t) = \nabla_\omega F^2(t, \omega_t).$$

Proof. Let $F = F^1 - F^2 \in \mathbb{C}^{1,1}_l(\Lambda_T)$ and $\omega \in C^0([0,T], \mathbb{R}^d)$. Then $F(t, \omega) = 0$ for all $t \leq T$. It is then obvious that $\mathcal{D}F(t, \omega)$ is also 0 on continuous paths. Assume now that there exists some $\omega \in C^0([0,T], \mathbb{R}^d)$ such that, for some $1 \leq i \leq d$ and $t_0 \in [0,T)$, $\partial_i F(t_0, \omega_{t_0}) > 0$. Let $\alpha = \partial_i F(t_0, \omega_{t_0})/2$. By the left-continuity of $\partial_i F$ and, using the fact that $\mathcal{D}F \in \mathbb{B}(\Lambda_T)$, there exists $\epsilon > 0$ such that for any $(t', \omega') \in \Lambda_T$,

$$[t' < t_0, \; d_\infty((t_0, \omega), (t', \omega')) < \epsilon] \implies [\partial_i F(t', \omega') > \alpha \text{ and } |\mathcal{D}F(t', \omega')| < 1].$$

Choose $t < t_0$ such that $d_\infty(\omega_t, \omega_{t_0}) < \epsilon/2$, put $h := t_0 - t$, define the following extension of ω_t to $[0, T]$,

$$z(u) = \omega(u), \quad u \leq t,$$
$$z_j(u) = \omega_j(t) + 1_{i=j}(u - t), \quad t \leq u \leq T, \quad 1 \leq j \leq d,$$

and define the following sequence of piecewise constant approximations of z_{t+h}:

$$z^n(u) = \tilde{z}^n = z(u) \quad \text{for} \quad u \leq t,$$

$$z_j^n(u) = \omega_j(t) + 1_{i=j} \frac{h}{n} \sum_{k=0}^{n} 1_{\frac{kh}{n} \leq u - t}, \quad \text{for} \quad t \leq u \leq t + h, \quad 1 \leq j \leq d.$$

Since $\|z_{t+h} - z_{t+h}^n\|_\infty = \frac{h}{n} \to 0$,

$$\left| F(t + h, z_{t+h}) - F(t + h, z_{t+h}^n) \right| \xrightarrow{n \to +\infty} 0.$$

We can now decompose $F(t + h, z_{t+h}^n) - F(t, \omega)$ as

$$F\left(t + h, z_{t+h}^n\right) - F(t, \omega) = \sum_{k=1}^{n} \left(F\left(t + \frac{kh}{n}, z_{t+\frac{kh}{n}}^n\right) - F\left(t + \frac{kh}{n}, z_{t+\frac{kh}{n}-}^n\right) \right)$$

$$+ \sum_{k=1}^{n} \left(F\left(t + \frac{kh}{n}, z_{t+\frac{kh}{n}-}^n\right) - F\left(t + \frac{(k-1)h}{n}, z_{t+\frac{(k-1)h}{n}}^n\right) \right),$$

where the first sum corresponds to jumps of z^n at times $t + kh/n$ and the second sum to the 'horizontal' variations of z^n on $[t + (k-1)h/n, t + kh/n]$. Write

$$F\left(t + \frac{kh}{n}, z_{t+\frac{kh}{n}}^n\right) - F\left(t + \frac{kh}{n}, z_{t+\frac{kh}{n}-}^n\right) = \phi\left(\frac{h}{n}\right) - \phi(0),$$

where

$$\phi(u) = F\left(t + \frac{kh}{n}, z_{t+\frac{kh}{n}-}^n + ue_i 1_{[t+\frac{kh}{n},T]}\right).$$

Since F is vertically differentiable, ϕ is differentiable and

$$\phi'(u) = \partial_i F\left(t + \frac{kh}{n}, z_{t+\frac{kh}{n}-}^n + ue_i 1_{[t+\frac{kh}{n},T]}\right)$$

is continuous. For $u \leq h/n$ we have

$$d_\infty\left((t, \omega_t), \left(t + \frac{kh}{n}, z_{(t+kh/n)-}^n + ue_i 1_{[t+\frac{kh}{n},T]}\right)\right) \leq h$$

so, $\phi'(u) > \alpha$ hence,

$$\sum_{k=1}^{n} F\left(t + \frac{kh}{n}, z_{t+\frac{kh}{n}}^n\right) - F\left(t + \frac{kh}{n}, z_{t+\frac{kh}{n}-}^n\right) > \alpha h.$$

On the other hand write

$$F\left(t + \frac{kh}{n}, z^n_{t + \frac{kh}{n}-}\right) - F\left(t + \frac{(k-1)h}{n}, z^n_{t+ \frac{(k-1)h}{n}}\right) = \psi\left(\frac{h}{n}\right) - \psi(0),$$

where

$$\psi(u) = F\left(t + \frac{(k-1)h}{n} + u, z^n_{t+ \frac{(k-1)h}{n}}\right).$$

So, ψ is right-differentiable on $]0, h/n[$ with right derivative

$$\psi'_r(u) = \mathcal{D}F\left(t + \frac{(k-1)h}{n} + u, z^n_{t+ \frac{(k-1)h}{n}}\right).$$

Since $F \in \mathbb{C}^{1,1}_l(\Lambda_T)$, ψ is left-continuous by Proposition 5.1.5, so

$$\sum_{k=1}^{n} \left(F\left(t + \frac{kh}{n}, z^n_{t+ \frac{kh}{n}-}\right) - F\left(t + \frac{(k-1)h}{n}, z^n_{t+ \frac{(k-1)h}{n}}\right)\right) = \int_0^h \mathcal{D}F\left(t + u, z^n_t\right) du.$$

Noting that

$$d_\infty\left((t + u, z^n_{t+u}), (t + u, z_{t+u})\right) \leq \frac{h}{n},$$

we obtain

$$\mathcal{D}F\left(t + u, z^n_{t+u}\right) \xrightarrow{n \to +\infty} \mathcal{D}F\left(t + u, z_{t+u}\right) = 0,$$

since the path of z_{t+u} is continuous. Moreover, $|\mathcal{D}F(t + u, z^n_{t+u})| \leq 1$ since $d_\infty((t + u, z^n_{t+u}), (t_0, \omega)) \leq \epsilon$ so, by dominated convergence, the integral converges to 0 as $n \to \infty$. Writing

$$F(t+h, z_{t+h}) - F(t, \omega) = [F(t+h, z_{t+h}) - F(t+h, z^n_{t+h})] + [F(t+h, z^n_{t+h}) - F(t, \omega)],$$

and taking the limit on $n \to \infty$ leads to $F(t + h, z_{t+h}) - F(t, \omega) \geq \alpha h$, a contradiction. $\qquad \square$

The above result implies in particular that, if $\nabla_\omega F^i \in \mathbb{C}^{1,1}(\Lambda_T)$, $\mathcal{D}(\nabla_\omega F) \in \mathbb{B}(\Lambda_T)$, and $F^1(\omega) = F^2(\omega)$ for any continuous path ω, then $\nabla^2_\omega F^1$ and $\nabla^2_\omega F^2$ must also coincide on continuous paths. The next theorem shows that this result can be obtained under the weaker assumption that $F^i \in \mathbb{C}^{1,2}(\Lambda_T)$, using a probabilistic argument. Interestingly, while the uniqueness of the first vertical derivative (Theorem 5.4.1) is based on the fundamental theorem of calculus, the proof of the following theorem is based on its stochastic equivalent, the Itô formula [40, 41].

Theorem 5.4.2. *If $F^1, F^2 \in \mathbb{C}^{1,2}_b(\Lambda_T)$ coincide on continuous paths, i.e.,*

$$\forall \omega \in C^0([0, T], \mathbb{R}^d), \quad \forall t \in [0, T), \quad F^1(t, \omega_t) = F^2(t, \omega_t),$$

then their second vertical derivatives also coincide on continuous paths:

$$\forall \omega \in C^0([0, T], \mathbb{R}^d), \quad \forall t \in [0, T), \quad \nabla^2_\omega F^1(t, \omega_t) = \nabla^2_\omega F^2(t, \omega_t).$$

Proof. Let $F = F^1 - F^2$. Assume that there exists $\omega \in C^0([0,T], \mathbb{R}^d)$ such that, for some $1 \leq i \leq d$ and $t_0 \in [0,T)$, and for some direction $h \in \mathbb{R}^d$, $\|h\| = 1$, we have ${}^th \cdot \nabla^2_\omega F(t_0, \omega_{t_0}) \cdot h > 0$, and denote $\alpha = ({}^th \cdot \nabla^2_\omega F(t_0, \omega_{t_0}) \cdot h)/2$. We will show that this leads to a contradiction. We already know that $\nabla_\omega F(t, \omega_t) = 0$ by Theorem 5.4.1. There exists $\eta > 0$ such that

$$\forall (t', \omega') \in \Lambda_T, \ \{t' \leq t_0, \quad d_\infty((t_0, \omega), (t', \omega')) < \eta\} \implies \tag{5.20}$$

$$\max\left(|F(t', \omega') - F(t_0, \omega_{t_0})|, \ |\nabla_\omega F(t', \omega')|, \ |DF(t', \omega')|\right) < 1, \ {}^th\nabla^2_\omega F(t', \omega')h > \alpha.$$

Choose $t < t_0$ such that $d_\infty(\omega_t, \omega_{t_0}) < \eta/2$, and denote $\epsilon = \frac{\eta}{2} \wedge (t_0 - t)$. Let W be a real Brownian motion on an (auxiliary) probability space $(\tilde{\Omega}, \mathcal{B}, \mathbb{P})$ whose generic element we will denote w, $(\mathcal{B}_s)_{s \geq 0}$ its natural filtration, and let

$$\tau = \inf\left\{s > 0, \ |W(s)| = \frac{\epsilon}{2}\right\}.$$

Define, for $t' \in [0,T]$, the 'Brownian extrapolation'

$$U_{t'}(\omega) = \omega(t')1_{t' \leq t} + \left(\omega(t) + W((t' - t) \wedge \tau)h\right)1_{t' > t}.$$

For all $s < \epsilon/2$, we have

$$d_\infty\left((t + s, U_{t+s}(\omega)), (t, \omega_t)\right) < \epsilon, \quad \mathbb{P}\text{-a.s.} \tag{5.21}$$

Define the following piecewise constant approximation of the stopped process W_τ:

$$W^n(s) = \sum_{i=0}^{n-1} W\left(i\frac{\epsilon}{2n} \wedge \tau\right)1_{s \in [i\frac{\epsilon}{2n}, (i+1)\frac{\epsilon}{2n})} + W\left(\frac{\epsilon}{2} \wedge \tau\right)1_{s=\frac{\epsilon}{2}}, \quad 0 \leq s \leq \frac{\epsilon}{2}.$$

Denoting

$$Z(s) = F(t + s, U_{t+s}), \quad s \in [0, T - t], \quad Z^n(s) = F(t + s, U^n_{t+s}), \tag{5.22}$$

$$U^n_{t'}(\omega) = \omega(t')1_{t' \leq t} + \left(\omega(t) + W^n((t' - t) \wedge \tau)h\right)1_{t' > t},$$

we have the following decomposition:

$$Z\left(\frac{\epsilon}{2}\right) - Z(0) = Z\left(\frac{\epsilon}{2}\right) - Z^n\left(\frac{\epsilon}{2}\right) + \sum_{i=1}^{n}\left(Z^n\left(i\frac{\epsilon}{2n}\right) - Z^n\left(i\frac{\epsilon}{2n}-\right)\right)$$

$$+ \sum_{i=0}^{n-1}\left(Z^n\left((i+1)\frac{\epsilon}{2n}-\right) - Z^n\left(i\frac{\epsilon}{2n}\right)\right). \tag{5.23}$$

The first term in (5.23) vanishes almost surely, since

$$\left\|U_{t+\frac{\epsilon}{2}} - U^n_{t+\frac{\epsilon}{2}}\right\|_\infty \xrightarrow{n \to \infty} 0.$$

The second term in (5.23) may be expressed as

$$Z^n\left(i\frac{\epsilon}{2n}\right) - Z^n\left(i\frac{\epsilon}{2n}-\right) = \phi_i\left(W\left(i\frac{\epsilon}{2n}\right) - W\left((i-1)\frac{\epsilon}{2n}\right)\right) - \phi_i(0), \qquad (5.24)$$

where

$$\phi_i(u,\omega) = F\left(t + i\frac{\epsilon}{2n}, U^n_{t+i\frac{\epsilon}{2n}-}(\omega) + uh1_{[t+i\frac{\epsilon}{2n},T]}\right).$$

Note that $\phi_i(u,\omega)$ is measurable with respect to $\mathcal{B}_{(i-1)\epsilon/2n}$, whereas (5.24) is independent with respect to $\mathcal{B}_{(i-1)\epsilon/2n}$. Let $\Omega_1 \subset \tilde{\Omega}$ and $\mathbb{P}(\Omega_1) = 1$ such that W has continuous sample paths on Ω_1. Then, on Ω_1, $\phi_i(\cdot,\omega) \in C^2(\mathbb{R})$ and the following relations hold $\mathbb{P}$-almost surely:

$$\phi_i'(u,\omega) = \nabla_\omega F\left(t + i\frac{\epsilon}{2n}, U^n_{t+i\frac{\epsilon}{2n}-}(\omega_t) + uh1_{[t+i\frac{\epsilon}{2n},T]}\right) \cdot h,$$

$$\phi_i''(u,\omega) = {}^t h\nabla_\omega^2 F\left(t + i\frac{\epsilon}{2n}, U^n_{t+i\frac{\epsilon}{2n}}, \omega_t) + uh1_{[t+i\frac{\epsilon}{2n},T]}\right) \cdot h.$$

So, using the above arguments, we can apply the Itô formula to (5.24) on Ω_1. Therefore, summing on i and denoting $i(s)$ the index such that $s \in [(i(s)-1)\epsilon/2n, i(s)\epsilon/2n)$, we obtain

$$\sum_{i=1}^{n}\left(Z^n\left(i\frac{\epsilon}{2n}\right) - Z^n\left(i\frac{\epsilon}{2n}-\right)\right)$$

$$= \int_0^{\frac{\epsilon}{2}} \nabla_\omega F\left(t + i(s)\frac{\epsilon}{2n}, U^n_{t+i(s)\frac{\epsilon}{2n}-}\right.$$

$$\left. + \left(W(s) - W\left((i(s)-1)\frac{\epsilon}{2n}\right)\right)h1_{[t+i(s)\frac{\epsilon}{2n},T]}\right)dW(s)$$

$$+ \frac{1}{2}\int_0^{\frac{\epsilon}{2}} \langle h, \nabla_\omega^2 F\left(t + i(s)\frac{\epsilon}{2n}, U^n_{t+i(s)\frac{\epsilon}{2n}-}\right.$$

$$\left. + \left(W(s) - W\left((i(s)-1)\frac{\epsilon}{2n}\right)\right)h1_{[t+i(s)\frac{\epsilon}{2n},T]}\right) \cdot h\rangle ds.$$

Since the first derivative is bounded by (5.20), the stochastic integral is a martingale so, taking expectation leads to

$$E\left[\sum_{i=1}^{n} Z^n\left(i\frac{\epsilon}{2n}\right) - Z^n\left(i\frac{\epsilon}{2n}-\right)\right] \geq \alpha\frac{\epsilon}{2}.$$

Write

$$Z^n\left((i+1)\frac{\epsilon}{2n}-\right) - Z^n\left(i\frac{\epsilon}{2n}\right) = \psi\left(\frac{\epsilon}{2n}\right) - \psi(0),$$

where

$$\psi(u) = F\left(t + i\frac{\epsilon}{2n} + u, U^n_{t+i\frac{\epsilon}{2n}}\right)$$

is right-differentiable with right derivative

$$\psi'(u) = \mathcal{D}F\left(t + i\frac{\epsilon}{2n} + u, \; U^n_{t+i\frac{\epsilon}{2n}}\right).$$

Since $F \in \mathbb{C}^{0,0}_l([0,T])$, ψ is left-continuous and the fundamental theorem of calculus yields

$$\sum_{i=0}^{n-1}\left(Z^n\left((i+1)\frac{\epsilon}{2n} -\right) - Z^n\left(i\frac{\epsilon}{2n}\right)\right) = \int_0^{\frac{\epsilon}{2}} \mathcal{D}F\left(t + s, U^n_{t+(i(s)-1)\frac{\epsilon}{2n}}\right)ds.$$

The integrand converges to $\mathcal{D}F(t+s, U_{t+s}) = 0$ as $n \to \infty$, since $\mathcal{D}F(t+s, \omega) = 0$ whenever ω is continuous. Since this term is also bounded, by dominated convergence, we have

$$\int_0^{\frac{\epsilon}{2}} \mathcal{D}F\left(t + s, U^n_{t+(i(s)-1)\frac{\epsilon}{2n}}\right)ds \; \overset{n\to\infty}{\longrightarrow} \; 0.$$

It is obvious that $Z(\epsilon/2) = 0$, since $F(t,\omega) = 0$ whenever ω is a continuous path. On the other hand, since all derivatives of F appearing in (5.23) are bounded, the dominated convergence theorem allows to take expectations of both sides in (5.23) with respect to the Wiener measure, and obtain $\alpha\epsilon/2 = 0$, a contradiction. □

Theorem 5.4.2 is a key result: it enables us to define the class $\mathbb{C}^{1,2}_b(\mathcal{W}_T)$ of non-anticipative functionals such that their restriction to $\mathcal{W}_T$ fulfills the conditions of Theorem 5.4.2, say

$$F \in \mathbb{C}^{1,2}_b(\mathcal{W}_T) \quad \Longleftrightarrow \quad \exists \tilde{F} \in \mathbb{C}^{1,2}_b(\Lambda_T), \quad \tilde{F}_{|\mathcal{W}_T} = F,$$

without having to extend the functional to the full space Λ_T.

For such functionals, coupling the proof of Theorem 5.3.6 with Theorem 5.4.2 yields the following result.

Theorem 5.4.3 (Pathwise change of variable formula for $\mathbb{C}^{1,2}_b(\mathcal{W}_T)$ functionals). *For any $F \in \mathbb{C}^{1,2}_b(\mathcal{W}_T)$ and $\omega \in C^0([0,T],\mathbb{R}^d) \cap Q^\pi([0,T],\mathbb{R}^d)$, the limit*

$$\int_0^T \nabla_\omega F(t,\omega_t)d^\pi \omega := \lim_{n\to\infty} \sum_{i=0}^{k(n)-1} \nabla_\omega F(t^n_i, \omega^n_{t^n_i}) \cdot (\omega(t^n_{i+1}) - \omega(t^n_i))$$

exists and

$$F(T,\omega_T) - F(0,\omega_0) =$$
$$\int_0^T \mathcal{D}F(t,\omega_t)dt + \int_0^T \nabla_\omega F(t,\omega_t)d^\pi \omega + \int_0^T \frac{1}{2}\mathrm{tr}\left({}^t\nabla^2_\omega F(t,\omega_t)d[\omega]\right).$$

5.5 Application to functionals of stochastic processes

Consider now a stochastic process $Z \colon [0,T] \times \Omega \to \mathbb{R}^d$ on $(\Omega, \mathcal{F}, (\mathcal{F}_t)_{t \in [0,T]}, \mathbb{P})$. The previous formula holds for functionals of Z along a sequence of partitions π on the set

$$\Omega_\pi(Z) = \{\omega \in \Omega, \quad Z(.,\omega) \in Q^\pi([0,T], \mathbb{R}^d)\}.$$

If we can construct a sequence of partitions π such that $\mathbb{P}(\Omega_\pi(Z)) = 1$, then the functional Itô formula will hold almost surely. Fortunately, this turns out be the case for many important classes of stochastic processes:

(i) Wiener process: if W is a Wiener process under $\mathbb{P}$, then, for any nested sequence of partitions π with $|\pi_n| \to 0$, the paths of W lie in $Q^\pi([0,T], \mathbb{R}^d)$ with probability 1 (see [47]):

$$\mathbb{P}(\Omega_\pi(W)) = 1 \quad \text{and} \quad \forall \omega \in \Omega_\pi(W), \big[W(\cdot,\omega)\big]_\pi(t) = t.$$

This is a classical result due to Lévy [47, Sec. 4, Theorem 5]. The nesting condition may be removed if one requires that $|\pi_n| \log n \to 0$, see [20].

(ii) Fractional Brownian motion: if B^H is a fractional Brownian motion with Hurst index $H \in (0.5, 1)$, then, for any sequence of partitions π with $n|\pi_n| \to 0$, the paths of B^H lie in $Q^\pi([0,T], \mathbb{R}^d)$ with probability 1 (see [20]):

$$\mathbb{P}(\Omega_\pi(B^H)) = 1 \quad \text{and} \quad \forall \omega \in \Omega_\pi(B^H), [B^H(\cdot,\omega)](t) = 0.$$

(iii) Brownian stochastic integrals: Let $\sigma \colon (\Lambda_T, d_\infty) \to \mathbb{R}$ be a Lipschitz map, B a Wiener process, and consider the Itô stochastic integral

$$X_\cdot = \int_0^\cdot \sigma(t, B_t) dB(t).$$

Then, for any sequence of partitions π with $|\pi_n| \log n \to 0$, the paths of X lie in $Q^\pi([0,T], \mathbb{R})$ with probability 1 and

$$[X](t,\omega) = \int_0^t \big|\sigma(u, B_u(\omega))\big|^2 du.$$

(iv) Lévy processes: if L is a Lévy process with triplet (b, A, ν), then for any sequence of partitions π with $n|\pi_n| \to 0$, $\mathbb{P}(\Omega_\pi(L)) = 1$ and

$$\forall \omega \in \Omega_\pi(L), \quad [L(\cdot,\omega)](t) = tA + \sum_{s \in [0,t]} \big|L(s,\omega) - L(s-,\omega)\big|^2.$$

Note that the only property needed for the change of variable formula to hold (and, thus, for the integral to exist pathwise) is the finite quadratic variation property, which does not require the process to be a semimartingale.

The construction of the Föllmer integral depends a priori on the sequence π of partitions. But in the case of a semimartingale, one can identify these limits of (non-anticipative) Riemann sums as Itô integrals, which guarantees that the limit is a.s. unique, independent of the choice of π. We now take a closer look at the semimartingale case.

Chapter 6

The functional Itô formula

6.1 Semimartingales and quadratic variation

We now consider a **semimartingale** X on a probability space $(\Omega, \mathcal{F}, \mathbb{P})$, equipped with the natural filtration $\mathbb{F} = (\mathcal{F}_t)_{t \geq 0}$ of X; X is a càdlàg process and the *stochastic integral*

$$\phi \in \mathbb{S}(\mathbb{F}) \longmapsto \int \phi \, dX$$

defines a functional on the set $\mathbb{S}(\mathbb{F})$ of simple $\mathbb{F}$-predictable processes, with the following continuity property: for any sequence $\phi^n \in \mathbb{S}(\mathbb{F})$ of simple predictable processes,

$$\sup_{[0,T] \times \Omega} |\phi^n(t, \omega) - \phi(t, \omega)| \overset{n \to \infty}{\longrightarrow} 0 \implies \int_0^{\cdot} \phi^n dX \overset{\mathbf{UCP}}{\underset{n \to \infty}{\longrightarrow}} \int_0^{\cdot} \phi \, dX,$$

where UCP stands for *uniform convergence in probability* on compact sets, see [62]. For any càglàd adapted process ϕ, the Itô integral $\int_0^{\cdot} \phi dX$ may be then constructed as a limit (in probability) of nonanticipative Riemann sums: for any sequence $(\pi_n)_{n \geq 1}$ of partitions of $[0, T]$ with $|\pi_n| \to 0$,

$$\sum_{\pi_n} \phi(t_k^n) \cdot (X(t_{k+1}^n) - X(t_k^n)) \overset{\mathbb{P}}{\underset{n \to \infty}{\longrightarrow}} \int_0^T \phi \, dX.$$

Let us recall some important properties of semimartingales (see [19, 62]):

(i) Quadratic variation: for any sequence of partitions $\pi = (\pi_n)_{n \geq 1}$ of $[0, T]$ with $|\pi_n| \to 0$ a.s,

$$\sum (X(t_{k+1}^n) - X(t_k^n))^2 \overset{\mathbb{P}}{\longrightarrow} [X]^T = [X]^c(T) + \sum_{0 \leq s \leq T} \Delta X(s)^2 < \infty;$$

(ii) Itô formula: for every $f \in C^2(\mathbb{R}^d, \mathbb{R})$,

$$f(X(t)) = f(X(0)) + \int_0^t \nabla f(X) dX + \int_0^t \frac{1}{2} \operatorname{tr}(\partial_{xx}^2 f(X) d[X]^c)$$

$$+ \sum_{0 \leq s \leq t} \left(f(X(s-) + \Delta X(s)) - f(X(s-)) - \nabla f(X(s-)) \cdot \Delta X(s) \right);$$

(iii) semimartingale decomposition: X has a unique decomposition

$$X = M^{\mathrm{d}} + M^{\mathrm{c}} + A,$$

where M^{c} is a continuous $\mathbb{F}$-local martingale, M^{d} is a pure-jump-$\mathbb{F}$-local martingale, and A is a continuous $\mathbb{F}$-adapted finite variation process;

(iv) the increasing process $[X]$ has a unique decomposition $[X] = [M]^{\mathrm{d}} + [M]^{\mathrm{c}}$, where $[M]^{\mathrm{d}}(t) = \sum_{0 \leq s \leq t} \Delta X(s)^2$ and $[M]^{\mathrm{c}}$ is a continuous increasing $\mathbb{F}$-adapted process;

(v) if X has continuous paths, then it has a unique decomposition $X = M + A$, where M is a continuous $\mathbb{F}$-local martingale and A is a continuous $\mathbb{F}$-adapted process with finite variation.

These properties have several consequences. First, if $F \in \mathbb{C}_l^{0,0}(\Lambda_T)$, then the non-anticipative Riemann sums

$$\sum_{t_k^n \in \pi_n} F\big(t_k^n, X_{t_k^n -}\big) \cdot \big(X(t_{k+1}^n) - X(t_k^n)\big)$$

converge in probability to the Itô stochastic integral $\int F(t, X)dX$. So, by the a.s. uniqueness of the limit, the Föllmer integral constructed along any sequence of partitions $\pi = (\pi_n)_{n \geq 1}$ with $|\pi_n| \to 0$ almost surely coincides with the Itô integral. In particular, the limit is independent of the choice of partitions, so we omit in the sequel the dependence of the integrals on the sequence π.

The semimartingale property also enables one to construct a partition with respect to which the paths of the semimartingale have the finite quadratic variation property with probability 1.

Proposition 6.1.1. *Let S be a semimartingale on $(\Omega, \mathcal{F}, \mathbb{P}, \mathbb{F} = (\mathcal{F}_t)_{t \geq 0})$, $T > 0$. There exists a sequence of partitions $\pi = (\pi_n)_{n \geq 1}$ of $[0, T]$ with $|\pi_n| \to 0$, such that the paths of S lie in $Q^\pi([0, T], \mathbb{R}^d)$ with probability 1:*

$$\mathbb{P}\big(\{\omega \in \Omega, \quad S(\cdot, \omega) \in Q^\pi([0, T], \mathbb{R}^d)\}\big) = 1.$$

Proof. Consider the dyadic partition $t_k^n = kT/2^n$, $k = 0, \ldots, 2^n$. Since

$$\sum_{\pi_n} (S(t_{k+1}^n) - S(t_k^n))^2 \to [S]_T$$

in probability, there exists a subsequence $(\pi_n)_{n \geq 1}$ of partitions such that

$$\sum_{t_i \in \pi_n} \big(S(t_i^n) - S(t_{i+1}^n)\big)^2 \overset{n \to \infty}{\longrightarrow} [S]_T, \quad \mathbb{P}\text{-a.s.}$$

This subsequence achieves the result.   □

The notion of semimartingale in the literature usually refers to a real-valued process but one can extend this to vector-valued semimartingales, [43]. For an $\mathbb{R}^d$-valued semimartingale $X = (X^1, \ldots, X^d)$, the above properties should be understood in the vector sense, and the quadratic (co-)variation process is an S_d^+-valued process, defined by

$$\sum_{t_i^n \in \pi_n,\, t_i^n \le t} \left(X(t_i^n) - X(t_{i+1}^n)\right) \cdot{}^{\mathrm t}\left(X(t_i^n) - X(t_{i+1}^n)\right) \xrightarrow[n \to \infty]{\mathbb{P}} [X](t),$$

where $t \to [X](t)$ is a.s. increasing in the sense of the order on positive symmetric matrices:

$$\forall t \ge 0, \quad \forall h > 0, \quad [X](t + h) - [X](t) \in S_d^+.$$

6.2 The functional Itô formula

Using Proposition 6.1.1, we can now apply the pathwise change of variable formula derived in Theorem 5.3.6 to any semimartingale. The following *functional Itô formula*, shown in [8, Proposition 6], is a consequence of Theorem 5.3.6 combined with Proposition 6.1.1.

Theorem 6.2.1 (Functional Itô formula: càdlàg case). *Let X be an $\mathbb{R}^d$-valued semimartingale and denote, for $t > 0$, $X_{t-}(u) = X(u)1_{[0,t[}(u) + X(t-)1_{[t,T]}(u)$. For any $F \in \mathbb{C}_{\mathrm{loc}}^{1,2}(\Lambda_T)$, and $t \in [0, T[$,*

$$F(t, X_t) - F(0, X_0) = \int_0^t \mathcal{D}F(u, X_u)du \tag{6.1}$$

$$+ \int_0^t \nabla_\omega F(u, X_u) dX(u) + \int_0^t \frac{1}{2}\mathrm{tr}\left(\nabla_\omega^2 F(u, X_u)\ d[X](u)\right)$$

$$+ \sum_{u \in]0,t]} \left[F(u, X_u) - F(u, X_{u-}) - \nabla_\omega F(u, X_{u-}) \cdot \Delta X(u)\right] \quad a.s.$$

In particular, $Y(t) = F(t, X_t)$ is a semimartingale: the class of semimartingales is stable under transformations by $\mathbb{C}_{\mathrm{loc}}^{1,2}(\Lambda_T)$ functionals.

More precisely, we can choose $\pi = (\pi_n)_{n \ge 1}$ with

$$\sum_{\pi_n} \left(X(t_i^n) - X(t_{i+1}^n)\right)^{\mathrm t}\left(X(t_i^n) - X(t_{i+1}^n)\right) \xrightarrow{n \to \infty} [X](t) \quad a.s.$$

So, setting

$$\Omega_X = \Big\{\omega \in \Omega, \ \sum_{\pi_n} \left(X(t_i) - X(t_{i+1})\right)^{\mathrm t}\left(X(t_i) - X(t_{i+1})\right) \xrightarrow{n \to \infty} [X](T) < \infty\Big\},$$

we have that $\mathbb{P}(\Omega_X) = 1$ and, for any $F \in \mathbb{C}^{1,2}_{\mathrm{loc}}(\Lambda_T)$ and any $\omega \in \Omega_X$, the limit

$$\int_0^t \nabla_\omega F\big(u, X_u(\omega)\big) d^\pi X(\omega) := \lim_{n \to \infty} \sum_{t_i^n \in \pi_n} \nabla_\omega F\big(t_i^n, X_{t_i^n}^n(\omega)\big) \cdot \big(X(t_{i+1}^n, \omega) - X(t_i^n, \omega)\big)$$

exists and, for all $t \in [0, T]$,

$$F(t, X_t(\omega)) - F(0, X_0(\omega)) = \int_0^t \mathcal{D}F(u, X_u(\omega))du + \int_0^t \nabla_\omega F(u, X_u(\omega))dX(\omega)$$

$$+ \int_0^t \frac{1}{2}\mathrm{tr}\left(\nabla_\omega^2 F(u, X_u(\omega))\ d[X(\omega)]\right)$$

$$+ \sum_{u \in]0, T]} \Big[F(u, X_u(\omega)) - F(u, X_{u-}(\omega))$$

$$- \nabla_\omega F(u, X_{u-}(\omega)) \cdot \Delta X(\omega)(u)\Big].$$

Remark 6.2.2. Note that, unlike the usual statement of the Itô formula (see, e.g., [62, Ch. II, Sect. 7]), the statement here is that there exists a set Ω_X on which the equality (6.1) holds pathwise for *any* $F \in \mathbb{C}^{1,2}_b([0, T])$. This is particularly useful when one needs to take a supremum over such F, such as in optimal control problems, since the null set does not depend on F.

In the continuous case, the functional Itô formula reduces to the following theorem from [8, 9, 21], which we give here for the sake of completeness.

Theorem 6.2.3 (Functional Itô formula: continuous case [8, 9, 21]). *Let X be a continuous semimartingale and $F \in \mathbb{C}^{1,2}_{\mathrm{loc}}(\mathcal{W}_T)$. For any $t \in [0, T[$,*

$$F(t, X_t) - F(0, X_0) = \int_0^t \mathcal{D}F(u, X_u)du \tag{6.2}$$

$$+ \int_0^t \nabla_\omega F(u, X_u)dX(u) + \int_0^t \frac{1}{2}\mathrm{tr}\left({}^t\nabla_\omega^2 F(u, X_u)d[X]\right) \ a.s.$$

In particular, $Y(t) = F(t, X_t)$ is a continuous semimartingale.

If $F(t, X_t) = f(t, X(t))$, where $f \in C^{1,2}([0, T] \times \mathbb{R}^d)$, this reduces to the standard Itô formula.

Note that, using Theorem 5.4.2, it is sufficient to require that $F \in \mathbb{C}^{1,2}_{\mathrm{loc}}(\mathcal{W}_T)$, rather than $F \in \mathbb{C}^{1,2}_{\mathrm{loc}}(\Lambda_T)$.

Theorem 6.2.3 shows that, for a continuous semimartingale X, any smooth non-anticipative functional $Y = F(X)$ depends on F and its derivatives only via their values on continuous paths, i.e., on $\mathcal{W}_T \subset \Lambda_T$. Thus, $Y = F(X)$ can be reconstructed from the "second order jet" $(\nabla_\omega F, \mathcal{D}F, \nabla_\omega^2 F)$ of the functional F on $\mathcal{W}_T$.

Although these formulas are implied by the stronger pathwise formula (Theorem 5.3.6), one can also give a direct probabilistic proof using the Itô formula [11]. We outline the main ideas of the probabilistic proof, which shows the role played by the different assumptions. A more detailed version may be found in [11].

Sketch of proof of Theorem 6.2.3. Consider first a càdlàg piecewise constant process

$$X(t) = \sum_{k=1}^{n} 1_{[t_k, t_{k+1}[}(t)\Phi_k,$$

where Φ_k are $\mathcal{F}_{t_k}$-measurable bounded random variables. Each path of X is a sequence of horizontal and vertical moves:

$$X_{t_{k+1}} = X_{t_k} + (\Phi_{k+1} - \Phi_k)1_{[t_{k+1}, T]}.$$

We now decompose each increment of F into a horizontal and vertical part:

$$
\begin{aligned}
F(t_{k+1}, X_{t_{k+1}}) - F(t_k, X_{t_k}) &= F(t_{k+1}, X_{t_{k+1}}) - F(t_{k+1}, X_{t_k}) && \text{(vertical)} \\
&+ F(t_{k+1}, X_{t_k}) - F(t_k, X_{t_k}) && \text{(horizontal)}
\end{aligned}
$$

The horizontal increment is the increment of $\phi(h) = F(t_k + h, X_{t_k})$. The fundamental theorem of calculus applied to ϕ yields

$$F(t_{k+1}, X_{t_k}) - F(t_k, X_{t_k}) = \phi(t_{k+1} - t_k) - \phi(0) = \int_{t_k}^{t_{k+1}} \mathcal{D}F(t, X_t)dt.$$

To compute the vertical increment, we apply the Itô formula to $\psi \in C^2(\mathbb{R}^d)$ defined by $\psi(u) = F(t_{k+1}, X_{t_k} + u1_{[t_{k+1}, T]})$. This yields

$$
\begin{aligned}
F(t_{k+1}, X_{t_{k+1}}) - F(t_{k+1}, X_{t_k}) &= \psi(X(t_{k+1}) - X(t_k)) - \psi(0) \\
&= \int_{t_k}^{t_{k+1}} \nabla_\omega F(t, X_t)dX(t) + \frac{1}{2}\int_{t_k}^{t_{k+1}} \text{tr}(\nabla_\omega^2 F(t, X_t)d[X]).
\end{aligned}
$$

Consider now the case of a continuous semimartingale X; the piecewise constant approximation $_nX$ of X along the partition π_n approximates X almost surely in supremum norm. Thus, by the previous argument, we have

$$F(T, {_nX_T}) - F(0, X_0) = \int_0^T \mathcal{D}F(t, {_nX_t})dt + \int_0^T \nabla_\omega F(t, {_nX_t})d_nX$$

$$+ \frac{1}{2}\int_0^T \text{tr}\left({}^t\nabla_\omega^2 F(_nX_t)d[_nX]\right).$$

Since $F \in \mathbb{C}_b^{1,2}(\Lambda_T)$, all derivatives involved in the expression are left-continuous in the d_∞ metric, which allows to control their convergence as $n \to \infty$. Using the local boundedness assumption $\nabla_\omega F, \mathcal{D}F, \nabla_\omega^2 F \in \mathbb{B}(\Lambda_T)$, we can then use the dominated convergence theorem, its extension to stochastic integrals [62, Ch. IV Theorem 32], and Lemma 5.3.7 to conclude that the Riemann–Stieltjes integrals converge almost surely, and the stochastic integral in probability, to the terms appearing in (6.2) as $n \to \infty$, see [11].   □

6.3   Functionals with dependence on quadratic variation

The results outlined in the previous sections all assume that the functionals involved, and their directional derivatives, are continuous in supremum norm. This is a severe restriction for applications in stochastic analysis, where functionals may involve quantities such as quadratic variation, which cannot be constructed as a continuous functional for the supremum norm.

More generally, in many applications such as statistics of processes, physics or mathematical finance, one is led to consider path dependent functionals of a semimartingale X and its quadratic variation process $[X]$ such as

$$\int_0^t g(t, X_t)d[X](t), \quad G(t, X_t, [X]_t) \quad \text{or} \quad E\big[G(T, X(T), [X](T))|\mathcal{F}_t\big],$$

where $X(t)$ denotes the value at time t and $X_t = (X(u), u \in [0, t])$ the path up to time t.

In Chapter 7 we will develop a weak functional calculus capable of handling all such examples in a weak sense. However, disposing of a pathwise interpretation is of essence in most applications and this interpretation is lost when passing to weak derivatives. In this chapter, we show how the pathwise calculus outlined in Chapters 5 and 6 can be extended to functionals depending on quadratic variation, while retaining the pathwise interpretation.

Consider a continuous $\mathbb{R}^d$-valued semimartingale with absolutely continuous quadratic variation,

$$[X](t) = \int_0^t A(u)du,$$

where A is an S_d^+-valued process. Denote by $\mathcal{F}_t$ the natural filtration of X.

The idea is to 'double the variables' and to represent the above examples in the form

$$Y(t) = F\big(t, \{X(u), 0 \le u \le t\}, \{A(u), 0 \le u \le t\}\big) = F(t, X_t, A_t), \qquad (6.3)$$

where $F\colon [0,T] \times D\big([0,T],\mathbb{R}^d\big) \times D\big([0,T], S_d^+\big) \to \mathbb{R}$ is a non-anticipative functional defined on an enlarged domain $D([0,T],\mathbb{R}^d) \times D([0,T], S_d^+)$, and represents the dependence of Y on the stopped path $X_t = \{X(u), 0 \le u \le t\}$ of X and its quadratic variation.

Introducing the process A as an additional variable may seem redundant: indeed $A(t)$ is itself $\mathcal{F}_t$-measurable, i.e., a functional of X_t. However, it is not a *continuous* functional on (Λ_T, d_∞). Introducing A_t as a second argument in the functional will allow us to control the regularity of Y with respect to $[X](t) = \int_0^t A(u)du$ simply by requiring continuity of F with respect to the "lifted process" (X, A). This idea is analogous to the approach of rough path theory [34, 35, 49], in which one controls integral functionals of a path in terms of the path jointly with its 'Lévy area'. Here, $d[X]$ is the symmetric part of the Lévy area, while the asymmetric part does not intervene in the change of variable formula. However, unlike the rough path construction, in our construction we do not need to resort to p-variation norms.

An important property in the above examples is that their dependence on the process A is either through $[X] = \int_0^\cdot A(u)du$, or through an integral functional of the form $\int_0^\cdot \phi d[X] = \int_0^\cdot \phi(u)A(u)du$. In both cases they satisfy the condition

$$F(t, X_t, A_t) = F(t, X_t, A_{t-}),$$

where, as before, we denote

$$\omega_{t-} = \omega 1_{[0,t[} + \omega(t-)1_{[t,T]}.$$

Following these ideas, we define the space $(\mathcal{S}_T, d_\infty)$ of stopped paths

$$\mathcal{S}_T = \left\{ (t, x(t \wedge \cdot), v(t \wedge \cdot)), (t, x, v) \in [0, T] \times D([0, T], \mathbb{R}^d) \times D([0, T], S_d^+) \right\},$$

and we will assume throughout this section that

Assumption 6.3.1. $F \colon (\mathcal{S}_T, d_\infty) \to \mathbb{R}$ is a non-anticipative functional with "predictable" dependence with respect to the second argument:

$$\forall (t, x, v) \in \mathcal{S}_T, \quad F(t, x, v) = F(t, x_t, v_{t-}). \tag{6.4}$$

The regularity concepts introduced in Chapter 5 carry out without any modification to the case of functionals on the larger space $\mathcal{S}_T$; we define the corresponding class of regular functionals $\mathbb{C}_{loc}^{1,2}(\mathcal{S}_T)$ by analogy with Definition 5.2.10. Condition (6.4) entails that vertical derivatives with respect to the variable v are zero, so the vertical derivative on the product space coincides with the vertical derivative with respect to the variable x; we continue to denote it as ∇_ω.

As the examples below show, the decoupling of the variables (X, A) leads to a much larger class of smooth functionals.

Example 6.3.1 (Integrals with respect to the quadratic variation). A process

$$Y(t) = \int_0^t g(X(u))d[X](u),$$

where $g \in C^0(\mathbb{R}^d)$, may be represented by the functional

$$F(t, x_t, v_t) = \int_0^t g(x(u))v(u)du.$$

Here, F verifies Assumption 6.3.1, $F \in \mathbb{C}_b^{1,\infty}$, with $\mathcal{D}F(t, x_t, v_t) = g(x(t))v(t)$ and $\nabla_\omega^j F(t, x_t, v_t) = 0$.

Example 6.3.2. The process $Y(t) = X(t)^2 - [X](t)$ is represented by the functional

$$F(t, x_t, v_t) = x(t)^2 - \int_0^t v(u)du. \tag{6.5}$$

Here, F verifies Assumption 6.3.1, and $F \in \mathbb{C}_b^{1,\infty}(\mathcal{S}_T)$, with $\mathcal{D}F(t, x, v) = -v(t)$, $\nabla_\omega F(t, x_t, v_t) = 2x(t)$, $\nabla_\omega^2 F(t, x_t, v_t) = 2$, and $\nabla_\omega^j F(t, x_t, v_t) = 0$ for $j \geq 3$.

Example 6.3.3. The process $Y = \exp(X - [X]/2)$ may be represented by $Y(t) = F(t, X_t, A_t)$, with

$$F(t, x_t, v_t) = \exp\left(x(t) - \frac{1}{2}\int_0^t v(u)du\right).$$

Elementary computations show that $F \in \mathbb{C}_b^{1,\infty}(\mathcal{S}_T)$ with

$$\mathcal{D}F(t, x, v) = -\frac{1}{2}v(t)F(t, x, v)$$

and $\nabla_\omega^j F(t, x_t, v_t) = F(t, x_t, v_t)$.

We can now state a change of variable formula for non-anticipative functionals which are allowed to depend on the path of X *and* its quadratic variation.

Theorem 6.3.4. *Let $F \in \mathbb{C}_b^{1,2}(\mathcal{S}_T)$ be a non-anticipative functional verifying (6.4). Then, for $t \in [0, T[$,*

$$F(t, X_t, A_t) - F_0(X_0, A_0) = \int_0^t \mathcal{D}_u F(X_u, A_u)du + \int_0^t \nabla_\omega F(u, X_u, A_u)dX(u)$$

$$+ \int_0^t \frac{1}{2}\mathrm{tr}\left(\nabla_\omega^2 F(u, X_u, A_u)\, d[X]\right) \quad a.s. \tag{6.6}$$

In particular, for any $F \in \mathbb{C}_b^{1,2}(\mathcal{S}_T)$, $Y(t) = F(t, X_t, A_t)$ is a continuous semimartingale.

Importantly, (6.6) contains no extra term involving the dependence on A, as compared to (6.2). As will be clear in the proof, due to the predictable dependence of F in A, one can in fact 'freeze' the variable A when computing the increment/differential of F so, locally, A acts as a parameter.

Proof of Theorem 6.3.4. Let us first assume that X does not exit a compact set K, and that $\|A\|_\infty \leq R$ for some $R > 0$. Let us introduce a sequence of random partitions $(\tau_k^n, k = 0, \ldots, k(n))$ of $[0, t]$, by adding the jump times of A to the dyadic partition $(t_i^n = it/2^n, i = 0, \ldots, 2^n)$:

$$\tau_0^n = 0, \quad \tau_k^n = \inf\left\{s > \tau_{k-1}^n \,\Big|\, \frac{2^n s}{t} \in \mathbb{N} \text{ or } |A(s) - A(s-)| > \frac{1}{n}\right\} \wedge t.$$

The construction of $(\tau_k^n, \ k = 0, \ldots, k(n))$ ensures that

$$\eta_n = \sup\left\{ |A(u) - A(\tau_i^n)| + |X(u) - X(\tau_{i+1}^n)| + \frac{t}{2^n}, \ i \le 2^n, \ u \in [\tau_i^n, \ \tau_{i+1}^n[\right\} \overset{n\to\infty}{\longrightarrow} 0.$$

Set $_nX = \sum_{i=0}^{\infty} X(\tau_{i+1}^n) 1_{[\tau_i^n, \tau_{i+1}^n)} + X(t) 1_{\{t\}}$, which is a càdlàg piecewise constant approximation of X_t, and $_nA = \sum_{i=0}^{\infty} A(\tau_i^n) 1_{[\tau_i^n, \tau_{i+1}^n)} + A(t) 1_{\{t\}}$, which is an adapted càdlàg piecewise constant approximation of A_t. Write $h_i^n = \tau_{i+1}^n - \tau_i^n$. Start with the decomposition

$$\begin{aligned}
F\big(&\tau_{i+1}^n, {}_nX_{\tau_{i+1}^n-}, {}_nA_{\tau_{i+1}^n-}\big) - F\big(\tau_i^n, {}_nX_{\tau_i^n-}, {}_nA_{\tau_i^n-}\big) \\
&= F\big(\tau_{i+1}^n, {}_nX_{\tau_{i+1}^n-}, {}_nA_{\tau_i^n}\big) - F\big(\tau_i^n, {}_nX_{\tau_i^n}, {}_nA_{\tau_i^n}\big) \\
&\quad + F\big(\tau_i^n, {}_nX_{\tau_i^n}, {}_nA_{\tau_i^n}-\big) - F\big(\tau_i^n, {}_nX_{\tau_i^n}-, {}_nA_{\tau_i^n}-\big),
\end{aligned} \tag{6.7}$$

where we have used (6.4) to deduce $F(\tau_i^n, {}_nX_{\tau_i^n}, {}_nA_{\tau_i^n}) = F(\tau_i^n, {}_nX_{\tau_i^n}, {}_nA_{\tau_i^n}-)$. The first term in (6.7) can be written $\psi(h_i^n) - \psi(0)$, where

$$\psi(u) = F(\tau_i^n + u, {}_nX_{\tau_i^n}, {}_nA_{\tau_i^n}).$$

Since $F \in \mathbb{C}_b^{1,2}(\mathcal{S}_T)$, ψ is right-differentiable and left-continuous by Proposition 5.1.5, so

$$F\big(\tau_{i+1}^n, {}_nX_{\tau_i^n}, {}_nA_{\tau_i^n}\big) - F\big(\tau_i^n, {}_nX_{\tau_i^n}, {}_nA_{\tau_i^n}\big) = \int_0^{\tau_{i+1}^n - \tau_i^n} \mathcal{D}F\big(\tau_i^n + u, {}_nX_{\tau_i^n}, {}_nA_{\tau_i^n}\big) du.$$

The second term in (6.7) can be written $\phi(X(\tau_{i+1}^n) - X(\tau_i^n)) - \phi(0)$, where $\phi(u) = F(\tau_i^n, {}_nX_{\tau_i^n}^u-, {}_nA_{\tau_i^n})$. Since $F \in \mathbb{C}_b^{1,2}$, $\phi \in C^2(\mathbb{R}^d)$ and

$$\nabla \phi(u) = \nabla_\omega F\big(\tau_i^n, {}_nX_{\tau_i^n}^u-, {}_nA_{\tau_i^n}\big), \quad \nabla^2 \phi(u) = \nabla_\omega^2 F\big(\tau_i^n, {}_nX_{\tau_i^n}^u-, {}_nA_{\tau_i^n}\big).$$

Applying the Itô formula to $\phi(X(\tau_i^n + s) - X(\tau_i^n))$ yields

$$\begin{aligned}
\phi\big(X(\tau_{i+1}^n) - X(\tau_i^n)\big) - \phi(0) &= \int_{\tau_i^n}^{\tau_{i+1}^n} \nabla_\omega F\big(\tau_i^n, {}_nX_{\tau_i^n-}^{X(s)-X(\tau_i^n)}, {}_nA_{\tau_i^n}\big) dX(s) \\
&\quad + \frac{1}{2} \int_{\tau_i^n}^{\tau_{i+1}^n} \text{tr}\left[{}^t\nabla_\omega^2 F\big(\tau_i^n, {}_nX_{\tau_i^n-}^{X(s)-X(\tau_i^n)}, {}_nA_{\tau_i^n}\big) d[X](s) \right].
\end{aligned}$$

Summing over $i \ge 0$ and denoting $i(s)$ the index such that $s \in [\tau_{i(s)}^n, \tau_{i(s)+1}^n[$, we have shown that

$$\begin{aligned}
F\big(t, {}_nX_t, {}_nA_t\big) - F\big(0, X_0, A_0\big) &= \int_0^t \mathcal{D}F\big(s, {}_nX_{\tau_{i(s)}^n}, {}_nA_{\tau_{i(s)}^n}\big) ds \\
&\quad + \int_0^t \nabla_\omega F\big(\tau_{i(s)}^n, {}_nX_{\tau_{i(s)}^n-}^{X(s)-X(\tau_{i(s)}^n)}, {}_nA_{\tau_{i(s)}^n}\big) dX(s) \\
&\quad + \frac{1}{2} \int_0^t \text{tr}\left[\nabla_\omega^2 F\big(\tau_{i(s)}^n, {}_nX_{\tau_{i(s)}^n-}^{X(s)-X(\tau_{i(s)}^n)}, {}_nA_{\tau_{i(s)}^n}\big) d[X](s) \right].
\end{aligned}$$

Now, $F(t, {}_nX_t, {}_nA_t)$ converges to $F(t, X_t, A_t)$ almost surely. Since all approximations of (X, A) appearing in the various integrals have a d_∞-distance from (X_s, A_s)
less than $\eta_n \to 0$, the continuity at fixed times of $\mathcal{D}F$ and the left-continuity of
$\nabla_\omega F$, $\nabla_\omega^2 F$ imply that the integrands appearing in the above integrals converge,
respectively, to $\mathcal{D}F(s, X_s, A_s), \nabla_\omega F(s, X_s, A_s), \nabla_\omega^2 F(s, X_s, A_s)$ as $n \to \infty$. Since
the derivatives are in $\mathbb{B}$, the integrands in the various above integrals are bounded
by a constant depending only on F, K, R and t, but not on s nor on ω. The dominated convergence and the dominated convergence theorem for the stochastic
integrals [62, Ch.IV Theorem 32] then ensure that the Lebesgue–Stieltjes integrals
converge almost surely, and the stochastic integral in probability, to the terms
appearing in (6.6) as $n \to \infty$.

Consider now the general case, where X and A may be unbounded. Let K_n
be an increasing sequence of compact sets with $\bigcup_{n \geq 0} K_n = \mathbb{R}^d$, and denote the
stopping times by

$$\tau_n = \inf \left\{ s < t : X(s) \notin K^n \text{ or } |A(s)| > n \right\} \wedge t.$$

Applying the previous result to the stopped process $(X_{t \wedge \tau_n}, A_{t \wedge \tau_n})$ and noting
that, by (6.4), $F(t, X_t, A_t) = F(t, X_t, A_{t-})$ leads to

$$F\left(t, X_{t \wedge \tau_n}, A_{t \wedge \tau_n}\right) - F(0, X_0, A_0) = \int_0^{t \wedge \tau_n} \mathcal{D}F(u, X_u, A_u)du$$

$$+ \frac{1}{2} \int_0^{t \wedge \tau_n} \text{tr}\left({}^t\nabla_\omega^2 F(u, X_u, A_u)d[X](u)\right)$$

$$+ \int_0^{t \wedge \tau_n} \nabla_\omega F(u, X_u, A_u)dX$$

$$+ \int_{t \wedge \tau^n}^t \mathcal{D}F\left(u, X_{u \wedge \tau_n}, A_{u \wedge \tau_n}\right)du.$$

The terms in the first line converges almost surely to the integral up to time t
since $t \wedge \tau_n = t$ almost surely for n sufficiently large. For the same reason the last
term converges almost surely to 0. $\square$

Chapter 7

Weak functional calculus for square-integrable processes

The pathwise functional calculus presented in Chapters 5 and 6 extends the Itô Calculus to a large class of path dependent functionals of semimartingales, of which we have already given several examples. Although in Chapter 6 we introduced a probability measure $\mathbb{P}$ on the space $D([0, T], \mathbb{R}^d)$, under which X is a semimartingale, its probabilistic properties did not play any crucial role in the results, since the key results were shown to hold pathwise, $\mathbb{P}$-almost surely.

However, this pathwise calculus requires continuity of the functionals and their directional derivatives with respect to the supremum norm: without the (left) continuity condition (5.1.4), the Functional Itô formula (Theorem 6.2.3) may fail to hold in a pathwise sense.[1] This excludes some important examples of non-anticipative functionals, such as Itô stochastic integrals or the Itô map [48, 52], which describes the correspondence between the solution of a stochastic differential equation and the driving Brownian motion.

Another issue which arises when trying to apply the pathwise functional calculus to stochastic processes is that, in a probabilistic setting, functionals of a process X need only to be defined on a set of probability 1 with respect to $\mathbb{P}^X$: modifying the definition of a functional F outside the support of $\mathbb{P}^X$ does not affect the image process $F(X)$ from a probabilistic perspective. So, in order to work with processes, we need to extend the previous framework to functionals which are not necessarily defined on the whole space Λ_T, but only $\mathbb{P}^X$-almost everywhere, i.e., $\mathbb{P}^X$-equivalence classes of functionals.

This construction is well known in a finite-dimensional setting: given a reference measure μ on $\mathbb{R}^d$, one can define the notions of μ-almost everywhere regularity, weak derivatives, and Sobolev regularity scales using duality relations with respect to the reference measure; this is the approach behind Schwartz's theory of distributions [65]. The Malliavin Calculus [52], conceived by Paul Malliavin as a weak calculus for Wiener functionals, can be seen as an analogue of Schwartz distribution theory on the Wiener space, using the Wiener measure as reference measure.[2] The idea is to construct an extension of the derivative operator using a

---

[1] Failure to recognize this important point has lead to several erroneous assertions in the literature, via a naive application of the functional Itô formula.

[2] This viewpoint on the Malliavin Calculus was developed by Shigekawa [66], Watanabe [73],

duality relation, which extends the *integration by parts* formula. The closure of the derivative operator then defines the class of function(al)s to which the derivative may be applied in a weak sense.

We adopt here a similar approach, outlined in [11]: given a square integrable Itô process X, we extend the pathwise functional calculus to a *weak nonanticipative functional calculus* whose domain of applicability includes all square integrable semimartingales adapted to the filtration $\mathcal{F}^X$ generated by X.

We first define, in Section 7.1, a vertical derivative operator acting on *processes*, i.e., $\mathbb{P}$-equivalence classes of functionals of X, which is shown to be related to the martingale representation theorem (Section 7.2).

This operator ∇_X is then shown to be closable on the space of square integrable martingales, and its closure is identified as the inverse of the Itô integral with respect to X: for $\phi \in \mathcal{L}^2(X)$, $\nabla_X \left(\int \phi dX \right) = \phi$ (Theorem 7.3.3). In particular, we obtain a constructive version of the martingale representation theorem (Theorem 7.3.4) stating that, for any square integrable $\mathcal{F}_t^X$-martingale Y,

$$Y(T) = Y(0) + \int_0^T \nabla_X Y \, dX \quad \mathbb{P}\text{-a.s.}$$

This formula can be seen as a non-anticipative counterpart of the Clark–Haussmann–Ocone formula [5, 36, 37, 44, 56]. The integrand $\nabla_X Y$ is an adapted process which may be computed pathwise, so this formula is more amenable to numerical computations than those based on Malliavin Calculus [12].

Section 7.4 further discusses the relation with the Mallavian derivative on the Wiener space. We show that the weak derivative ∇_X may be viewed as a non-anticipative "lifting" of the Malliavin derivative (Theorem 7.4.1): for square integrable martingales Y whose terminal value $Y(T) \in \mathbf{D}^{1,2}$ is Malliavin differentiable, we show that the vertical derivative is in fact a version of the predictable projection of the Malliavin derivative, $\nabla_X Y(t) = E[\mathbb{D}_t H | \mathcal{F}_t]$.

Finally, in Section 7.5 we extend the weak derivative to all square integrable semimartingales and discuss an application of this construction to forward-backward SDEs (FBSDEs, for short).

7.1   Vertical derivative of an adapted process

Throughout this section, $X \colon [0, T] \times \Omega \to \mathbb{R}^d$ will denote a continuous, $\mathbb{R}^d$-valued semimartingale defined on a probability space $(\Omega, \mathcal{F}, \mathbb{P})$. Since all processes we deal with are functionals of X, we will consider, without loss of generality, Ω to be the canonical space $D([0, T], \mathbb{R}^d)$, and $X(t, \omega) = \omega(t)$ to be the coordinate process. Then, $\mathbb{P}$ denotes the law of the semimartingale. We denote by $\mathbb{F} = (\mathcal{F}_t)_{t \geq 0}$ the $\mathbb{P}$-completion of $\mathcal{F}_{t+}$.

---

and Sugita [70, 71].

We assume that

$$[X](t) = \int_0^t A(s)ds \qquad (7.1)$$

for some càdlàg process A with values in S_d^+. Note that A need not be a semi-martingale.

Any non-anticipative functional $F\colon \Lambda_T \to \mathbb{R}$ applied to X generates an $\mathbb{F}$-adapted process

$$Y(t) = F(t, X_t) = F\big(t, \{X(u \wedge t), u \in [0, T]\}\big). \qquad (7.2)$$

However, the functional representation of the process Y is not unique: modifying F outside the topological support of $\mathbb{P}^X$ does not change Y. In particular, the values taken by F outside $\mathcal{W}_T$ do not affect Y. Yet, the definition of the vertical $\nabla_\omega F$ seems to depend on the value of F for paths such as $\omega + e1_{[t,T]}$, which clearly do not belong to $\mathcal{W}_T$.

Theorem 5.4.2 gives a partial answer to this issue in the case where the topological support of $\mathbb{P}$ is the full space $C^0([0, T], \mathbb{R}^d)$ (which is the case, for instance, for the Wiener process), but does not cover the wide variety of situations that may arise. To tackle this issue we need to define a vertical derivative operator which acts on ($\mathbb{F}$-adapted) processes, i.e., equivalence classes of (non-anticipative) functionals modulo an evanescent set. The functional Itô formula (6.2) provides the key to this construction. If $F \in \mathbb{C}_{\text{loc}}^{1,2}(\mathcal{W}_T)$, then Theorem 6.2.3 identifies $\int_0^\cdot \nabla_\omega F(t, X_t)dM$ as the local martingale component of $Y(t) = F(t, X_t)$, which implies that it should be independent from the functional representation of Y.

Lemma 7.1.1. *Let $F^1, F^2 \in \mathbb{C}_b^{1,2}(\mathcal{W}_T)$ be such that, for every $t \in [0, T[$,*

$$F^1(t, X_t) = F^2(t, X_t) \quad \mathbb{P}\text{-a.s.}$$

Then,

$${}^t\big[\nabla_\omega F^1(t, X_t) - \nabla_\omega F^2(t, X_t)\big]A(t-)\big[\nabla_\omega F^1(t, X_t) - \nabla_\omega F^2(t, X_t)\big] = 0$$

outside an evanescent set.

Proof. Let $X = Z + M$, where Z is a continuous process with finite variation and M is a continuous local martingale. There exists $\Omega_1 \in \mathcal{F}$ such that $\mathbb{P}(\Omega_1) = 1$ and, for $\omega \in \Omega_1$, the path $t \mapsto X(t, \omega)$ is continuous and $t \mapsto A(t, \omega)$ is càdlàg. Theorem 6.2.3 implies that the local martingale part of $0 = F^1(t, X_t) - F^2(t, X_t)$ can be written

$$0 = \int_0^t \big(\nabla_\omega F^1(u, X_u) - \nabla_\omega F^2(u, X_u)\big)dM(u).$$

Computing its quadratic variation we have, on Ω_1,

$$0 = \int_0^t \frac{1}{2} {}^t \big[\nabla_\omega F^1(u, X_u) - \nabla_\omega F^2(u, X_u) \big] A(u-) \big[\nabla_\omega F^1(u, X_u) - \nabla_\omega F^2(u, X_u) \big] \, du$$

$$(7.3)$$

and $\nabla_\omega F^1(t, X_t) = \nabla_\omega F^1(t, X_{t-})$ since X is continuous. So, on Ω_1, the integrand in (7.3) is left-continuous; therefore (7.3) implies that, for $t \in [0, T[$ and $\omega \in \Omega_1$,

$${}^t \big[\nabla_\omega F^1(t, X_t) - \nabla_\omega F^2(t, X_t) \big] A(t-) \big[\nabla_\omega F^1(t, X_t) - \nabla_\omega F^2(t, X_t) \big] = 0. \qquad \square$$

$$\square$$

Thus, if we assume that in (7.1) $A(t-)$ is almost surely non-singular, then Lemma 7.1.1 allows to define intrinsically the pathwise derivative of any process Y with a smooth functional representation $Y(t) = F(t, X_t)$.

Assumption 7.1.1. [Non-degeneracy of local martingale component] $A(t)$ in (7.1) is non-singular almost everywhere, i.e.,

$$\det(A(t)) \neq 0 \quad dt \times d\mathbb{P}\text{-a.e.}$$

Definition 7.1.2 (Vertical derivative of a process). *Define* $\mathcal{C}_{\mathrm{loc}}^{1,2}(X)$ *as the set of* $\mathcal{F}_t$-*adapted processes* Y *which admit a functional representation in* $\mathbb{C}_{\mathrm{loc}}^{1,2}(\Lambda_T)$:

$$\mathcal{C}_{\mathrm{loc}}^{1,2}(X) = \{ Y, \quad \exists F \in \mathbb{C}_{\mathrm{loc}}^{1,2}, \quad Y(t) = F(t, X_t) \quad dt \times \mathbb{P}\text{-a.e.} \}. \qquad (7.4)$$

Under Assumption 7.1.1, for any $Y \in \mathcal{C}_{\mathrm{loc}}^{1,2}(X)$, *the predictable process*

$$\nabla_X Y(t) = \nabla_\omega F(t, X_t)$$

is uniquely defined up to an evanescent set, independently of the choice of $F \in$ $\mathbb{C}_{\mathrm{loc}}^{1,2}(\mathcal{W}_T)$ *in the representation* (7.4). *We will call the process* $\nabla_X Y$ *the* vertical derivative *of* Y *with respect to* X.

Although the functional $\nabla_\omega F \colon \Lambda_T \to \mathbb{R}^d$ does depend on the choice of the functional representation F in (7.2), the *process* $\nabla_X Y(t) = \nabla_\omega F(t, X_t)$ obtained by computing $\nabla_\omega F$ along the paths of X does not.

The vertical derivative ∇_X defines a correspondence between $\mathbb{F}$-adapted processes: it maps any process $Y \in \mathcal{C}_b^{1,2}(X)$ to an $\mathbb{F}$-adapted process $\nabla_X Y$. It defines a linear operator

$$\nabla_X \colon \mathcal{C}_b^{1,2}(X) \longrightarrow \mathcal{C}_l^{0,0}(X)$$

on the algebra of smooth functionals $\mathcal{C}_b^{1,2}(X)$, with the following properties: for any $Y, Z \in \mathcal{C}_b^{1,2}(X)$, and any $\mathbb{F}$-predictable process λ, we have

(i) $\mathcal{F}_t$-linearity: $\nabla_X(Y + \lambda Z)(t) = \nabla_X Y(t) + \lambda(t) \cdot \nabla_X Z(t)$;

(ii) differentiation of the product: $\nabla_X (YZ)(t) = Z(t)\nabla_X Y(t) + Y(t)\nabla_X Z(t)$;

(iii) composition rule: If $U \in \mathcal{C}_b^{1,2}(Y)$, then $U \in \mathcal{C}_b^{1,2}(X)$ and, for every $t \in [0, T]$,

$$\nabla_X U(t) = \nabla_Y U(t) \cdot \nabla_X Y(t). \tag{7.5}$$

The vertical derivative operator ∇_X has important links with the Itô stochastic integral. The starting point is the following proposition.

Proposition 7.1.3 (Representation of smooth local martingales). *For any local martingale* $Y \in \mathcal{C}_{\text{loc}}^{1,2}(X)$ *we have the representation*

$$Y(T) = Y(0) + \int_0^T \nabla_X Y \, dM, \tag{7.6}$$

where M is the local martingale component of X.

Proof. Since $Y \in \mathcal{C}_{\text{loc}}^{1,2}(X)$, there exists $F \in \mathbb{C}_{\text{loc}}^{1,2}(\mathcal{W}_T)$ such that $Y(t) = F(t, X_t)$. Theorem 6.2.3 then implies that, for $t \in [0, T)$,

$$Y(t) - Y(0) = \int_0^t \mathcal{D}F(u, X_u) du + \frac{1}{2}\int_0^t \text{tr}\big({}^t\nabla_\omega^2 F(u, X_u) d[X](u)\big)$$
$$+ \int_0^t \nabla_\omega F(u, X_u) dZ(u) + \int_0^t \nabla_\omega F(t, X_t) dM(u).$$

Given the regularity assumptions on F, all terms in this sum are continuous processes with finite variation, while the last term is a continuous local martingale. By uniqueness of the semimartingale decomposition of Y, we thus have $Y(t) = \int_0^t \nabla_\omega F(u, X_u) dM(u)$. Since $F \in \mathbb{C}_l^{0,0}([0, T])$, $Y(t) \to F(T, X_T)$ as $t \to T$ so, the stochastic integral is also well-defined for $t \in [0, T]$. □

We now explore further the consequences of this property and its link with the martingale representation theorem.

7.2 Martingale representation formula

Consider now the case where X is a Brownian martingale,

$$X(t) = X(0) + \int_0^t \sigma(u) dW(u),$$

where σ is a process adapted to $\mathcal{F}_t^W$ and satisfying

$$E\left(\int_0^T \|\sigma(t)\|^2 dt\right) < \infty \quad \text{and} \quad \det(\sigma(t)) \neq 0, \quad dt \times d\mathbb{P}\text{-a.e.}$$

Then X is a square integrable martingale with the predictable representation property [45, 63]: for any square integrable $\mathcal{F}_T$ measurable random variable H or, equivalently, any square integrable $\mathbb{F}$-martingale Y defined by $Y(t) = E[H|\mathcal{F}_t]$, there exists a unique $\mathbb{F}$-predictable process ϕ such that

$$Y(t) = Y(0) + \int_0^t \phi dX, \quad \text{i.e.,} \quad H = Y(0) + \int_0^T \phi dX, \tag{7.7}$$

and

$$E\left(\int_0^T \operatorname{tr}(\phi(u) \cdot {}^t\phi(u)d[X](u)) \right) < \infty. \tag{7.8}$$

The classical proof of this representation result (see, e.g., [63]) is non-constructive and a lot of effort has been devoted to obtaining an explicit form for ϕ in terms of Y, using Markovian methods [18, 25, 27, 42, 58] or Malliavin Calculus [3, 5, 37, 44, 55, 56]. We will now see that the vertical derivative gives a simple representation of the integrand ϕ.

If $Y \in \mathcal{C}_{\mathrm{loc}}^{1,2}(X)$ is a martingale, then Proposition 7.1.3 applies so, comparing (7.7) with (7.6) suggests $\phi = \nabla_X Y$ as the obvious candidate. We now show that this guess is indeed correct in the square integrable case.

Let $\mathcal{L}^2(X)$ be the Hilbert space of $\mathbb{F}$-predictable processes such that

$$||\phi||_{\mathcal{L}^2(X)}^2 = E\left(\int_0^T \operatorname{tr}(\phi(u) \cdot {}^t\phi(u)d[X](u)) \right) < \infty.$$

This is the space of integrands for which one defines the usual L^2 extension of the Itô integral with respect to X.

Proposition 7.2.1 (Representation of smooth L^2 martingales). *If $Y \in \mathcal{C}_{\mathrm{loc}}^{1,2}(X)$ is a square integrable martingale, then $\nabla_X Y \in \mathcal{L}^2(X)$ and*

$$Y(T) = Y(0) + \int_0^T \nabla_X Y dX. \tag{7.9}$$

Proof. Applying Proposition 7.1.3 to Y we obtain that $Y = Y(0) + \int_0 \nabla_X Y dX$. Since Y is square integrable, the Itô integral $\int_0 \nabla_X Y dX$ is square integrable and the Itô isometry formula yields

$$E\big(||Y(T) - Y(0)||^2\big) = E\left| \int_0^T \nabla_X Y dX \right|^2$$

$$= E\left(\int_0^T \operatorname{tr}\big({}^t\nabla_X Y(u) \cdot \nabla_X Y(u)d[X](u)\big) \right) < \infty,$$

so $\nabla_X Y \in \mathcal{L}^2(X)$. The uniqueness of the predictable representation in $\mathcal{L}^2(X)$ then allows us to conclude the proof. $\square$

7.3 Weak derivative for square integrable functionals

Let $\mathcal{M}^2(X)$ be the space of square integrable $\mathbb{F}$-martingales with initial value zero, and with the norm $||Y|| = \sqrt{E|Y(T)|^2}$.

We will now use Proposition 7.2.1 to extend ∇_X to a continuous functional on $\mathcal{M}^2(X)$.

First, observe that the predictable representation property implies that the stochastic integral with respect to X defines a bijective isometry

$$I_X \colon \mathcal{L}^2(X) \longrightarrow \mathcal{M}^2(X),$$

$$\phi \longmapsto \int_0^\cdot \phi dX.$$

Then, by Proposition 7.2.1, the vertical derivative ∇_X defines a continuous map

$$\nabla_X \colon \mathcal{C}_b^{1,2}(X) \cap \mathcal{M}^2(X) \longrightarrow \mathcal{L}^2(X)$$

on the set of 'smooth martingales'

$$D(X) = \mathcal{C}_b^{1,2}(X) \cap \mathcal{M}^2(X).$$

This isometry property can now be used to extend ∇_X to the entire space $\mathcal{M}^2(X)$, using the following density argument.

Lemma 7.3.1 (Density of $\mathcal{C}_b^{1,2}(X)$ in $\mathcal{M}^2(X)$). $\{\nabla_X Y \mid Y \in D(X)\}$ *is dense in* $\mathcal{L}^2(X)$ *and* $D(X) = \mathcal{C}_b^{1,2}(X) \cap \mathcal{M}^2(X)$ *is dense in* $\mathcal{M}^2(X)$.

Proof. We first observe that the set of cylindrical non-anticipative processes of the form

$$\phi_{n,f,(t_1,\ldots,t_n)}(t) = f\big(X(t_1),\ldots,X(t_n)\big)1_{t>t_n},$$

where $n \geq 1$, $0 \leq t_1 < \cdots < t_n \leq T$, and $f \in C_b^\infty(\mathbb{R}^n, \mathbb{R})$, is a total set in $\mathcal{L}^2(X)$, i.e., their linear span, which we denote by U, is dense in $\mathcal{L}^2(X)$. For such an integrand $\phi_{n,f,(t_1,\ldots,t_n)}$, the stochastic integral with respect to X is given by the martingale

$$Y(t) = I_X(\phi_{n,f,(t_1,\ldots,t_n)})(t) = F(t, X_t),$$

where the functional F is defined on Λ_T as

$$F(t,\omega) = f\big(\omega(t_1-),\ldots,\omega(t_n-)\big)(\omega(t) - \omega(t_n))1_{t>t_n}.$$

Hence,

$$\nabla_\omega F(t,\omega) = f\big(\omega(t_1-),\ldots,\omega(t_n-)\big)1_{t>t_n}, \quad \nabla_\omega^2 F(t,\omega) = 0, \quad \mathcal{D}F(t,\omega_t) = 0,$$

which shows that $F \in \mathbb{C}_b^{1,2}$ (see Example 5.2.5). Hence, $Y \in \mathcal{C}_b^{1,2}(X)$. Since f is bounded, Y is obviously square integrable so, $Y \in D(X)$. Hence, $I_X(U) \subset D(X)$.

Since I_X is a bijective isometry from $\mathcal{L}^2(X)$ to $\mathcal{M}^2(X)$, the density of U in $\mathcal{L}^2(X)$ entails the density of $I_X(U)$ in $\mathcal{M}^2(X)$. $\square$

This leads to the following useful characterization of the vertical derivative ∇_X.

Proposition 7.3.2 (Integration by parts on $D(X)$)**.** *Let* $Y \in D(X) = \mathcal{C}_b^{1,2}(X) \cap \mathcal{M}^2(X)$. *Then,* $\nabla_X Y$ *is the unique element from* $\mathcal{L}^2(X)$ *such that, for every* $Z \in D(X)$,

$$E\big(Y(T)Z(T)\big) = E\bigg(\int_0^T \operatorname{tr}\big(\nabla_X Y(t)^{\mathrm{t}} \nabla_X Z(t) d[X](t)\big)\bigg). \tag{7.10}$$

Proof. Let $Y, Z \in D(X)$. Then Y is a square integrable martingale with $Y(0) = 0$ and $E[|Y(T)|^2] < \infty$. Applying Proposition 7.2.1 to Y and Z, we obtain $Y = \int \nabla_X Y dX, Z = \int \nabla_X Z dX$, so

$$E\big[Y(T)Z(T)\big] = E\bigg[\int_0^T \nabla_X Y dX \int_0^T \nabla_X Z dX\bigg].$$

Applying the Itô isometry formula we get (7.10). If we now consider another process $\psi \in \mathcal{L}^2(X)$ such that, for every $Z \in D(X)$,

$$E\big(Y(T)Z(T)\big) = E\bigg(\int_0^T \operatorname{tr}\big(\psi(t)^{\mathrm{t}} \nabla_X Z(t) d[X]\big)\bigg),$$

then, substracting from (7.10), we get

$$\langle \psi - \nabla_X Y, \nabla_X Z \rangle_{\mathcal{L}^2(X)} = 0$$

for all $Z \in D(X)$. And this implies $\psi = \nabla_X Y$ since, by Lemma 7.3.1, $\{\nabla_X Z \mid Z \in D(X)\}$ is dense in $\mathcal{L}^2(X)$. $\qquad\square$

We can rewrite (7.10) as

$$E\bigg(Y(T) \int_0^T \phi \, dX\bigg) = E\bigg(\int_0^T \operatorname{tr}\big(\nabla_X Y(t)^{\mathrm{t}} \phi(t) d[X](t)\big)\bigg), \tag{7.11}$$

which can be understood as an 'integration by parts formula' on $[0,T] \times \Omega$ with respect to the measure $d[X] \times d\mathbb{P}$.

The integration by parts formula, and the density of the domain are the two ingredients needed to show that the operator is closable on $\mathcal{M}^2(X)$.

Theorem 7.3.3 (Extension of ∇_X to $\mathcal{M}^2(X)$)**.** *The vertical derivative*

$$\nabla_X : D(X) \longrightarrow \mathcal{L}^2(X)$$

admits a unique continuous extension to $\mathcal{M}^2(X)$, *namely*

$$\nabla_X : \mathcal{M}^2(X) \longrightarrow \mathcal{L}^2(X),$$

$$\int_0^{\cdot} \phi dX \longmapsto \phi, \tag{7.12}$$

which is a bijective isometry characterized by the integration by parts formula: *for* $Y \in \mathcal{M}^2(X)$, $\nabla_X Y$ *is the unique element of* $\mathcal{L}^2(X)$ *such that, for every* $Z \in D(X)$,

$$E[Y(T)Z(T)] = E\left(\int_0^T \mathrm{tr}(\nabla_X Y(t)^t \nabla_X Z(t)d[X](t)) \right). \qquad (7.13)$$

In particular, ∇_X *is the adjoint of the Itô stochastic integral*

$$I_X : \mathcal{L}^2(X) \longrightarrow \mathcal{M}^2(X),$$
$$\phi \longmapsto \phi dX$$

in the following sense: $\forall \phi \in \mathcal{L}^2(X)$, $\forall Y \in \mathcal{M}^2(X)$,

$$E[Y(T) \int_0^T \phi dX] = \langle \nabla_X Y, \phi \rangle_{\mathcal{L}^2(X)}.$$

Proof. Any $Y \in \mathcal{M}^2(X)$ can be written as $Y(t) = \int_0^t \phi(s)dX(s)$ with $\phi \in \mathcal{L}^2(X)$, which is uniquely defined $d[X] \times d\mathbb{P}$-a.e. Then, the Itô isometry formula guarantees that (7.13) holds for ϕ. To show that (7.13) uniquely characterizes ϕ, consider $\psi \in \mathcal{L}^2(X)$ also satisfying (7.13); then, denoting $I_X(\psi) = \int_0^{\cdot} \psi dX$ its stochastic integral with respect to X, (7.13) implies that, for every $Z \in D(X)$,

$$\langle I_X(\psi) - Y, Z \rangle_{\mathcal{M}^2(X)} = E\left[(Y(T) - \int_0^T \psi dX)Z(T) \right] = 0;$$

and this implies that $I_X(\psi) = Y$ $d[X] \times d\mathbb{P}$-a.e. since, by construction, $D(X)$ is dense in $\mathcal{M}^2(X)$. Hence, $\nabla_X : D(X) \mapsto \mathcal{L}^2(X)$ is closable on $\mathcal{M}^2(X)$. $\qquad \square$

We have thus extended Dupire's pathwise vertical derivative ∇_ω to a weak derivative ∇_X defined for all square integrable $\mathbb{F}$-martingales, which is the inverse of the Itô integral I_X with respect to X: for every $\phi \in \mathcal{L}^2(X)$, the equality

$$\nabla_X \left(\int_0^{\cdot} \phi \, dX \right) = \phi$$

holds in $\mathcal{L}^2(X)$.

The above results now allow us to state a general version of the martingale representation formula, valid for all square integrable martingales.

Theorem 7.3.4 (Martingale representation formula: general case). *For any square integrable* $\mathbb{F}$-*martingale* Y *and every* $t \in [0, T]$,

$$Y(t) = Y(0) + \int_0^t \nabla_X Y dX \quad \mathbb{P}\text{-}a.s.$$

This relation, which can be understood as a stochastic version of the 'fundamental theorem of calculus' for martingales, shows that ∇_X is a 'stochastic derivative' in the sense of Zabczyk [75] and Davis [18].

Theorem 7.3.3 suggests that the space of square integrable martingales can be seen as a 'Martingale Sobolev space' of order 1 constructed over $\mathcal{L}^2(X)$, see [11]. However, since for $Y \in \mathcal{M}^2(X), \nabla_X Y$ is not a martingale, Theorem 7.3.3 does not allow to iterate this weak differentiation to higher orders. We will see in Section 7.5 how to extend this construction to semimartingales and construct Sobolev spaces of arbitrary order for the operator ∇_X.

7.4   Relation with the Malliavin derivative

The above construction holds, in particular, in the case where $X = W$ is a Wiener process. Consider the canonical Wiener space $(\Omega_0 = C_0([0,T],\mathbb{R}^d), \|\cdot\|_\infty, \mathbb{P})$ endowed with the filtration of the canonical process W, which is a Wiener process under $\mathbb{P}$.

Theorem 7.3.3 applies when $X = W$ is the Wiener process and allows to define, for any square integrable Brownian martingale $Y \in \mathcal{M}^2(W)$, the *vertical derivative* $\nabla_W Y \in \mathcal{L}^2(W)$ of Y with respect to W.

Note that the Gaussian properties of W play no role in the construction of the operator ∇_W. We now compare the Brownian vertical derivative ∇_W with the *Malliavin derivative* [3, 4, 52, 67].

Consider an $\mathcal{F}_T$-measurable functional $H = H(X(t), t \in [0,T]) = H(X_T)$ with $E[|H|^2] < \infty$. If H is differentiable in the Malliavin sense [3, 52, 55, 67], e.g., $H \in \mathbf{D}^{1,2}$ with Malliavin derivative $\mathbb{D}_t H$, then the Clark–Haussmann–Ocone formula [55, 56] gives a stochastic integral representation of H in terms of the Malliavin derivative of H:

$$H = E[H] + \int_0^T {}^\mathrm{P} E\big[\mathbb{D}_t H \mid \mathcal{F}_t\big] dW(t), \qquad (7.14)$$

where ${}^\mathrm{P} E[\mathbb{D}_t H|\mathcal{F}_t]$ denotes the predictable projection of the Malliavin derivative. This yields a stochastic integral representation of the martingale $Y(t) = E[H|\mathcal{F}_t]$:

$$Y(t) = E[H \mid \mathcal{F}_t] = E[H] + \int_0^t {}^\mathrm{P} E\big[\mathbb{D}_t H \mid \mathcal{F}_u\big] dW(u).$$

Related martingale representations have been obtained under a variety of conditions [3, 18, 27, 44, 55, 58].

Denote by

- $L^2([0,T] \times \Omega)$ the set of $\mathcal{F}_T$-measurable maps $\phi\colon [0,T] \times \Omega \to \mathbb{R}^d$ on $[0,T]$ with

$$E \int_0^T \|\phi(t)\|^2 dt < \infty;$$

- $E[\cdot \,|\, \mathbb{F}]$ the conditional expectation operator with respect to the filtration $\mathbb{F} = (\mathcal{F}_t, t \in [0, T])$:

$$E[\cdot \,|\, \mathbb{F}] : H \in L^2(\Omega, \mathcal{F}_T, \mathbb{P}) \longrightarrow (E[H \,|\, \mathcal{F}_t], t \in [0, T]) \in \mathcal{M}^2(W);$$

- $\mathbb{D}$ the Malliavin derivative operator, which associates to a random variable $H \in \mathbf{D}^{1,2}(0, T)$ the (anticipative) process $(\mathbb{D}_t H)_{t \in [0,T]} \in L^2([0, T] \times \Omega)$.

Theorem 7.4.1 (Intertwining formula). *The following diagram is commutative in the sense of $dt \times d\mathbb{P}$ equalities:*

$$
\begin{array}{ccc}
\mathcal{M}^2(W) & \xrightarrow{\nabla_W} & \mathcal{L}^2(W) \\
{\scriptstyle E[.|\mathbb{F}]}\Big\uparrow & & \Big\uparrow{\scriptstyle E[.|\mathbb{F}]} \\
\mathbf{D}^{1,2} & \xrightarrow[\mathbb{D}]{} & L^2([0, T] \times \Omega)
\end{array}
$$

In other words, the conditional expectation operator intertwines ∇_W with the Malliavin derivative: for every $H \in \mathbf{D}^{1,2}(\Omega_0, \mathcal{F}_T, \mathbb{P})$,

$$\nabla_W \big(E[H \,|\, \mathcal{F}_t] \big) = E \big[\mathbb{D}_t H \,|\, \mathcal{F}_t \big]. \tag{7.15}$$

Proof. The Clark–Haussmann–Ocone formula [56] gives that, for every $H \in \mathbf{D}^{1,2}$,

$$H = E[H] + \int_0^T {}^{\mathrm{P}} E \big[\mathbb{D}_t H | \mathcal{F}_t \big] dW_t,$$

where ${}^{\mathrm{P}} E[\mathbb{D}_t H | \mathcal{F}_t]$ denotes the predictable projection of the Malliavin derivative. On the other hand, from Theorem 7.2.1 we know that

$$H = E[H] + \int_0^T \nabla_W Y(t) dW(t),$$

where $Y(t) = E[H | \mathcal{F}_t]$. Therefore, ${}^{\mathrm{P}} E[\mathbb{D}_t H | \mathcal{F}_t] = \nabla_W E[H | \mathcal{F}_t]$, $dt \times d\mathbb{P}$-almost everywhere. $\qquad \square$

The *predictable projection* on $\mathcal{F}_t$ (see [19, Vol. I]) can be viewed as a morphism which "lifts" relations obtained in the framework of Malliavin Calculus to relations between non-anticipative quantities, where the Malliavin derivative and the Skorokhod integral are replaced, respectively, by the vertical derivative ∇_W and the Itô stochastic integral.

Rewriting (7.15) as

$$(\nabla_W Y)(t) = E \big[\mathbb{D}_t(Y(T)) \,|\, \mathcal{F}_t \big]$$

for every $t \le T$, we can note that the choice of $T \ge t$ is arbitrary: for any $h \ge 0$, we have

$$\nabla_W Y(t) = {}^{\mathrm{P}} E \big[\mathbb{D}_t(Y(t+h)) \,|\, \mathcal{F}_t \big].$$

Taking the limit $h \to 0+$ yields

$$\nabla_W Y(t) = \lim_{h \to 0+} {}^{\mathrm{P}} E\big[\mathbb{D}_t Y(t+h) \,|\, \mathcal{F}_t\big].$$

The right-hand side is sometimes written, with some abuse of notation, as $\mathbb{D}_t Y(t)$, and called the 'diagonal Malliavin derivative'. From a computational viewpoint, unlike the Clark–Haussmann–Ocone representation, which requires to simulate the *anticipative* process $\mathbb{D}_t H$ and compute conditional expectations, $\nabla_X Y$ only involves non-anticipative quantities which can be computed path by path. It is thus more amenable to numerical computations, see [12].

	Malliavin derivative $\mathbb{D}$	Vertical Derivative ∇_W
Perturbations	$H^1([0,T], \mathbb{R}^d)$	$\{e\, 1_{[t,T]}, e \in \mathbb{R}^d, t \in [0,T]\}$
Domain	$\mathbf{D}^{1,2}(\mathcal{F}_T)$	$\mathcal{M}^2(W)$
Range	$L^2([0,T] \times \Omega)$	$\mathcal{L}^2(W)$
Measurability	$\mathcal{F}_T^W$-measurable (anticipative)	$\mathcal{F}_t^W$-measurable (non-anticipative)
Adjoint	Skorokhod integral	Itô integral

Table 7.1: Comparison of the Malliavin derivative and the vertical derivative.

The commutative diagram above shows that ∇_W may be seen as 'non-anticipative lifting' of the Malliavin derivative: the Malliavin derivative on random variables (lower level) is lifted to the vertical derivative of non-anticipative processes (upper level) by the conditional expectation operator.

Having pointed out the parallels between these constructions, we must note, however, that our construction of the weak derivative operator ∇_X works for any square integrable continuous martingale X, and does not involve any Gaussian or probabilistic properties of X. Also, the domain of the vertical derivative spans all square integrable functionals, whereas the Malliavin derivative has a strictly smaller domain $\mathbf{D}^{1,2}$.

Regarding this last point, the above correspondence *can* be extended to the full space $L^2(\mathbb{P}, \mathcal{F}_T)$ using Hida's *white noise analysis* [38] and the use of distribution-valued processes. Aase, Oksendal, and Di Nunno [1] extend the Malliavin derivative as a distribution-valued process $\widetilde{\mathbb{D}}_t F$ on $L^2(\mathbb{P}, \mathcal{F}_T)$, where $\mathbb{P}$ is the Wiener measure, $\mathcal{F}_T$ is the Brownian filtration, and derive the following generalized Clark–Haussman–Ocone formula: for $F \in L^2(\mathbb{P}, \mathcal{F}_T)$,

$$F = E[F] + \int_0^T E\big[\widetilde{\mathbb{D}}_t F \,|\, \mathcal{F}_t\big] \diamond W(t) dt,$$

where $\diamond$ denotes a 'Wick product' [1].

Although this white noise extension of the Malliavin derivative is not a stochastic process but a distribution-valued process, the uniqueness of the martingale representation shows that its predictable projection $E[\widetilde{\mathbb{D}_t F}|\mathcal{F}_t]$ is a bona-fide square integrable process which is a version of the (weak) vertical derivative:

$$E\big[\widetilde{\mathbb{D}_t F}\,|\,\mathcal{F}_t\big] = \nabla_W Y(t) \quad dt \times d\mathbb{P}\text{-a.e.}$$

This extends the commutative diagram to the full space

$$
\begin{array}{ccc}
\mathcal{M}^2(W) & \xrightarrow{\ \nabla_W\ } & \mathcal{L}^2(W) \\
{\scriptstyle (E[.|\mathcal{F}_t])_{t\in[0,T]}} \Big\uparrow & & \Big\uparrow {\scriptstyle (E[.|\mathcal{F}_t])_{t\in[0,T]}} \\
\mathbf{L}^2(\Omega, \mathcal{F}_T, \mathbb{P}) & \xrightarrow[\ \widetilde{\mathbb{D}}\ ]{} & \mathbf{H}^{-1}([0,T]\times\Omega)
\end{array}
$$

However, the formulas obtained using the vertical derivative ∇_W only involve Itô stochastic integrals of non-anticipative processes and do not require any recourse to distribution-valued processes.

But, most importantly, the construction of the vertical derivative and the associated martingale representation formulas make no use of the Gaussian properties of the underlying martingale X and extend, well beyond the Wiener process, to any square integrable martingale.

7.5 Extension to semimartingales

We now show how ∇_X can be extended to all square integrable $\mathbb{F}$-*semimartingales*.

Define $\mathcal{A}^2(\mathbb{F})$ as the set of continuous $\mathbb{F}$-predictable absolutely continuous processes $H = H(0) + \int_0^\cdot h(t)dt$ with finite variation such that

$$\|H\|_{\mathcal{A}^2}^2 = E|H(0)|^2 + \|h\|_{L^2(dt\times d\mathbb{P})}^2 = E\left(|H(0)|^2 + \int_0^T |h(t)|^2 dt\right) < \infty,$$

and consider the direct sum

$$\mathcal{S}^{1,2}(X) = \mathcal{M}^2(X) \oplus \mathcal{A}^2(\mathbb{F}). \qquad (7.16)$$

Then, any process $S \in \mathcal{S}^{1,2}(X)$ is an $\mathbb{F}$-adapted special semimartingale with a unique decomposition $S = M + H$, where $M \in \mathcal{M}^2(X)$ is a square integrable $\mathbb{F}$-martingale with $M(0) = 0$, and $H \in \mathcal{A}^2(\mathbb{F})$ with $H(0) = S(0)$. The norm given by

$$\|S\|_{1,2}^2 = E\left([S](T)\right) + \|H\|_{\mathcal{A}^2}^2 \qquad (7.17)$$

defines a Hilbert space structure on $\mathcal{S}^{1,2}(X)$ for which $\mathcal{A}^2(\mathbb{F})$ and $\mathcal{M}^2(X)$ are closed subspaces of $\mathcal{S}^{1,2}(X)$.

We will refer to processes in $\mathcal{S}^{1,2}(X)$ as 'square integrable $\mathbb{F}$-semimartingales'.

Theorem 7.5.1 (Weak derivative on $\mathcal{S}^{1,2}(X)$). *The vertical derivative*

$$\nabla_X : \mathcal{C}^{1,2}(X) \cap \mathcal{S}^{1,2}(X) \longrightarrow \mathcal{L}^2(X)$$

admits a unique continuous extension to the space $\mathcal{S}^{1,2}(X)$ of square integrable $\mathbb{F}$-semimartingales, such that:

(i) *the restriction of ∇_X to square integrable martingales is a bijective isometry*

$$\nabla_X : \mathcal{M}^2(X) \longrightarrow \mathcal{L}^2(X),$$
$$\int_0^{\cdot} \phi dX \longmapsto \phi \tag{7.18}$$

   which is the inverse of the Itô integral with respect to X;

(ii) *for any finite variation process $H \in \mathcal{A}^2(X)$, $\nabla_X H = 0$.*

   Clearly, both properties are necessary since ∇_X already verifies them on $\mathcal{S}^{1,2}(X) \cap \mathcal{C}_b^{1,2}(X)$, which is dense in $\mathcal{S}^{1,2}(X)$. Continuity of the extension then imposes $\nabla_X H = 0$ for any $H \in \mathcal{A}^2(\mathbb{F})$. Uniqueness results from the fact that (7.16) is a continuous direct sum of closed subspaces. The semimartingale decomposition of $S \in \mathcal{S}^{1,2}(X)$ may then be interpreted as a projection on the closed subspaces $\mathcal{A}^2(\mathbb{F})$ and $\mathcal{M}^2(X)$.

   The following characterization follows from the definition of $\mathcal{S}^{1,2}(X)$.

Proposition 7.5.2 (Description of $\mathcal{S}^{1,2}(X)$). *For every $S \in \mathcal{S}^{1,2}(X)$ there exists a unique $(h, \phi) \in \mathcal{L}^2(dt \times d\mathbb{P}) \times \mathcal{L}^2(X)$ such that*

$$S(t) = S(0) + \int_0^t h(u)du + \int_0^t \phi dX,$$

where $\phi = \nabla_X S$ and

$$h(t) = \frac{d}{dt}\left(S(t) - \int_0^t \nabla_X S dX\right).$$

   The map $\nabla_X : \mathcal{S}^{1,2}(X) \to \mathcal{L}^2(X)$ is continuous. So, $\mathcal{S}^{1,2}(X)$ may be seen as a Sobolev space of order 1 constructed above $\mathcal{L}^2(X)$, which explains our notation.

   We can iterate this construction and define a scale of 'Sobolev' spaces of $\mathbb{F}$-semimartingales with respect to ∇_X.

Theorem 7.5.3 (The space $\mathcal{S}^{k,2}(X)$). *Let $\mathcal{S}^{0,2}(X) = \mathcal{L}^2(X)$ and define, for $k \geq 2$,*

$$\mathcal{S}^{k,2}(X) := \left\{S \in \mathcal{S}^{1,2}(X), \quad \nabla_X S \in \mathcal{S}^{k-1,2}(X)\right\} \tag{7.19}$$

equipped with the norm

$$\|S\|_{k,2}^2 = \|H\|_{\mathcal{A}^2}^2 + \sum_{j=1}^k \|\nabla_X^j S\|_{\mathcal{L}^2(X)}^2.$$

Then $(\mathcal{S}^{k,2}(X), \|.\|_{k,2})$ *is a Hilbert space,* $\mathcal{S}^{k,2}(X) \cap \mathbb{C}_b^{1,2}(X)$ *is dense in* $\mathcal{S}^{k,2}(X)$, *and the maps*

$$\nabla_X : \mathcal{S}^{k,2}(X) \longrightarrow \mathcal{S}^{k-1,2}(X) \quad and \quad I_X : \mathcal{S}^{k-1,2}(X) \longrightarrow \mathcal{S}^{k,2}(X)$$

are continuous.

The above construction enables us to define

$$\nabla_X^k : \mathcal{S}^{k,2}(X) \longrightarrow \mathcal{L}^2(X)$$

as a continuous operator on $\mathcal{S}^{k,2}(X)$. This construction may be viewed as a non-anticipative counterpart of the Gaussian Sobolev spaces introduced in the Malliavin Calculus [26, 66, 70, 71, 73]; unlike those constructions, the embeddings involve the Itô stochastic integral and do not rely on the Gaussian nature of the underlying measure.

We can now use these ingredients to extend the horizontal derivative (Definition 5.2.1) to $S \in \mathcal{S}^{2,2}(X)$. First, note that if $S = F(X)$ with $F \in \mathbb{C}_b^{1,2}(\mathcal{W}_T)$, then

$$S(t) - S(0) - \int_0^t \nabla_X S dX - \frac{1}{2} \int_0^t \langle \nabla_X^2 S, d[X] \rangle \tag{7.20}$$

is equal to $\int_0^t \mathcal{D}F(u) du$. For $S \in \mathcal{S}^{2,2}(X)$, the process defined by (7.20) is almost surely absolutely continuous with respect to the Lebesgue measure: we use its Radon–Nikodým derivative to define a weak extension of the horizontal derivative.

Definition 7.5.4 (Extension of the horizontal derivative to $\mathcal{S}^{2,2}$). *For* $S \in \mathcal{S}^{2,2}(X)$ *there exists a unique* $\mathbb{F}$-*adapted process* $\mathcal{D}S \in \mathcal{L}^2(dt \times d\mathbb{P})$ *such that, for every* $t \in [0, T]$,

$$\int_0^t \mathcal{D}S(u) du = S(t) - S(0) - \int_0^t \nabla_X S dX - \frac{1}{2} \int_0^t \nabla_X^2 S d[X] \tag{7.21}$$

and

$$E \left(\int_0^T |\mathcal{D}S(t)|^2 dt \right) < \infty.$$

Equation (7.21) shows that $\mathcal{D}S$ may be interpreted as the 'Stratonovich drift' of S, i.e.,

$$\int_0^t \mathcal{D}S(u) du = S(t) - S(0) - \int_0^t \nabla_X S \circ dX,$$

where $\circ$ denotes the Stratonovich integral.

The following property is a straightforward consequence of Definition 7.5.4.

Proposition 7.5.5. *Let* $(Y^n)_{n \geq 1}$ *be a sequence in* $\mathcal{S}^{2,2}(X)$. *If* $\|Y^n - Y\|_{2,2} \xrightarrow{n \to \infty} 0$, *then* $\mathcal{D}Y^n(t) \to \mathcal{D}Y(t)$, $dt \times d\mathbb{P}$-*a.e.*

The above discussion implies that Proposition 7.5.5 does not hold if $\mathcal{S}^{2,2}(X)$ is replaced by $\mathcal{S}^{1,2}(X)$ (or $\mathcal{L}^2(X)$).

If $S = F(X)$ with $F \in \mathbb{C}^{1,2}_{\text{loc}}(\mathcal{W}_T)$, the Functional Itô formula (Theorem 6.2.3) then implies that $\mathcal{D}S$ is a version of the process $\mathcal{D}F(t, X_t)$. However, as the following example shows, Definition 7.5.4 is much more general and applies to functionals which are not even continuous in the supremum norm.

Example 7.5.6 (Multiple stochastic integrals). Let

$$(\Phi(t))_{t\in[0,T]} = (\Phi_1(t), \ldots, \Phi_d(t))_{t\in[0,T]}$$

be an $\mathbb{R}^{d\times d}$-valued $\mathbb{F}$-adapted process such that

$$E \int_0^T \sum_{i,j=1}^d \left(\int_0^t \text{tr}\big(\Phi_i(u)^t \Phi_j(u) d[X](u)\big) \right) d[X]_{i,j}(t) < \infty.$$

Then $\int_0^\cdot \Phi dX$ defines an $\mathbb{R}^d$-valued $\mathbb{F}$-adapted process whose components are in $\mathcal{L}^2(X)$. Consider now the process

$$Y(t) = \int_0^t \left(\int_0^s \Phi(u) dX(u) \right) dX(s). \tag{7.22}$$

Then $Y \in \mathcal{S}^{2,2}(X)$, $\nabla_X Y(t) = \int_0^t \Phi dX$, $\nabla^2_X Y(t) = \Phi(t)$, and

$$\mathcal{D}Y(t) = -\frac{1}{2}\langle \Phi(t), A(t) \rangle,$$

where $A(t) = d[X]/dt$ and $\langle \Phi, A \rangle = \text{tr}(^t\Phi.A)$.

Note that, in the example above, the matrix-valued process Φ need not be symmetric.

Example 7.5.7. In Example 7.5.6 (take $d = 2$), $X = (W^1, W^2)$ is a two-dimensional standard Wiener process with $\text{cov}(W^1(t), W^2(t)) = \rho t$. Let

$$\Phi(t) = \begin{pmatrix} 0 & 0 \\ 1 & 0 \end{pmatrix}.$$

Then

$$Y(t) = \int_0^t \left(\int_0^s \Phi dX \right) dX(s) = \int_0^t \begin{pmatrix} 0 \\ W^1 \end{pmatrix} dX = \int_0^t W^1 dW^2$$

verifies $Y \in \mathcal{S}^{2,2}(X)$, with

$$\nabla_X Y(t) = \int_0^t \Phi dX = \begin{pmatrix} 0 \\ W^1(t) \end{pmatrix}, \quad \nabla^2_X Y(t) = \Phi = \begin{pmatrix} 0 & 0 \\ 1 & 0 \end{pmatrix}, \quad \mathcal{D}Y = -\frac{\rho}{2}.$$

In particular, $\nabla_X^2 Y$ is not symmetric. Note that a naive pathwise calculation of the horizontal derivative gives '$\mathcal{D}Y = 0$', which is incorrect: the correct calculation is based on (7.21).

This example should be contrasted with the situation for pathwise-differentiable functionals: as discussed in Subsection 5.2.3, for $F \in \mathbb{C}_b^{1,2}(\Lambda_T)$, $\nabla_\omega^2 F(t, \omega)$ is a symmetric matrix. This implies that Y has no smooth functional representation, i.e., $Y \notin \mathcal{C}_b^{1,2}(X)$.

Example 7.5.6 is fundamental and can be used to construct in a similar manner martingale elements of $\mathcal{S}^{k,2}(X)$ for any $k \geq 2$, by using multiple stochastic integrals of square integrable tensor-valued processes. The same reasoning as above can then be used to show that, for any square integrable *non-symmetric k-tensor* process, its multiple (k-th order) stochastic integral with respect to X yields an element of $\mathcal{S}^{k,2}(X)$ which does not belong to $\mathcal{C}_b^{1,k}(X)$.

The following example clarifies some incorrect remarks found in the literature.

Example 7.5.8 (Quadratic variation). The quadratic variation process $Y = [X]$ is a square integrable functional of X: $Y \in \mathcal{S}^{2,2}(X)$ and, from Definition 7.5.4 we obtain $\nabla_X Y = 0$, $\nabla_X^2 Y = 0$ and $\mathcal{D}Y(t) = A(t)$, where $A(t) = d[X]/dt$.

Indeed, since $[X]$ has finite variation: $Y \in \ker(\nabla_X)$. This is consistent with the pathwise approach used in [11] (see Section 6.3), which yields the same result.

These ingredients now allow us to state a weak form of the Functional Itô formula, *without any pathwise-differentiability* requirements.

Proposition 7.5.9 (Functional Itô formula: weak form). *For any semimartingale $S \in \mathcal{S}^{2,2}(X)$, the following equality holds $dt \times d\mathbb{P}$-a.e.:*

$$S(t) = S(0) + \int_0^t \nabla_X S dX + \int_0^t \mathcal{D}S(u)du + \frac{1}{2}\int_0^t \text{tr}\left({}^t\nabla_X^2 S d[X]\right). \qquad (7.23)$$

For $S \in \mathcal{C}_{\text{loc}}^{1,2}(X)$ the terms in the formula coincide with the pathwise derivatives in Theorem 6.2.3. The requirement $S \in \mathcal{S}^{2,2}(X)$ is necessary for defining all terms in (7.23) as square integrable processes: although one can compute a weak derivative $\nabla_X S$ for $S \in \mathcal{S}^{1,2}(X)$, if the condition $S \in \mathcal{S}^{2,2}(X)$ is removed, in general the terms in the decomposition of $S - \int \nabla_X S dX$ are distribution-valued processes.

7.6 Changing the reference martingale

Up to now we have considered a fixed reference process X, assumed in Sections 7.2–7.5 to be a square integrable martingale verifying Assumption 7.1.1. One of the strong points of this construction is precisely that the choice of the reference martingale X is arbitrary; unlike the Malliavin Calculus, we are not restricted to choosing X to be Gaussian. We will now show that the operator ∇_X transforms in a simple way under a change of the reference martingale X.

Proposition 7.6.1 (*'Change of reference martingale'*). *Let X be a square integrable Itô martingale verifying Assumption 7.1.1. If $M \in \mathcal{M}^2(X)$ and $S \in \mathcal{S}^{1,2}(M)$, then $S \in \mathcal{S}^{1,2}(X)$ and*

$$\nabla_X S(t) = \nabla_M S(t) \cdot \nabla_X M(t) \quad dt \times d\mathbb{P}\text{-}a.e.$$

Proof. Since $M \in \mathcal{M}^2(X)$, by Theorem 7.3.4, $M = \int_0^{\cdot} \nabla_X M dX$ and

$$[M](t) = \int_0^t \operatorname{tr}\left(\nabla_X M \cdot^t \nabla_X M d[X]\right).$$

But $\mathcal{F}_t^M \subset \mathcal{F}_t^X$ so, $\mathcal{A}^2(M) \subset \mathcal{A}^2(\mathbb{F})$. Since $S \in \mathcal{S}^{1,2}(M)$, there exists $H \in \mathcal{A}^2(M) \subset \mathcal{A}^2(\mathbb{F})$ and $\nabla_M S \in \mathcal{L}^2(M)$ such that

$$S = H + \int_0^{\cdot} \nabla_M S dM = H + \int_0^{\cdot} \nabla_M S \nabla_X M dX$$

and

$$E\left(\int_0^T |\nabla_M S(t)|^2 \operatorname{tr}\left(\nabla_X M \cdot^t \nabla_X M d[X]\right)\right) =$$

$$E\left(\int_0^T |\nabla_M S|^2 d[M]\right) = \|\nabla_M S\|_{\mathcal{L}^2(M)}^2 < \infty.$$

Hence, $S \in \mathcal{S}^{1,2}(X)$. Uniqueness of the semimartingale decomposition of S then entails $\int_0 \nabla_M S \nabla_X M dX = \int_0 \nabla_X S dX$. Using Assumption 7.1.1 and following the same steps as in the proof of Lemma 7.1.1, we conclude that $\nabla_X S(t) = \nabla_M S(t) \cdot \nabla_X M(t)$ $dt \times d\mathbb{P}$-a.e. $\qquad \square$

7.7   Forward-Backward SDEs

Consider now a forward-backward stochastic differential equation (FBSDE) (see, e.g., [57, 58]) driven by a continuous semimartingale X:

$$X(t) = X(0) + \int_0^t b(u, X_{u-}, X(u))du + \int_0^t \sigma(u, X_{u-}, X(u))dW(u), \qquad (7.24)$$

$$Y(t) = H(X_T) - \int_t^T f(s, X_{s-}, X(s), Y(s), Z(s))ds - \int_t^T Z dX, \qquad (7.25)$$

where we have distinguished in our notation the dependence with respect to the path X_{t-} on $[0, t[$ from the dependence with respect to its endpoint $X(t)$, i.e., the stopped path (t, X_t) is represented as $(t, X_{t-}, X(t))$. Here, X is called the *forward* process, $\mathbb{F} = (\mathcal{F}_t)_{t \geq 0}$ is the $\mathbb{P}$-completed natural filtration of X, $H \in L^2(\Omega, \mathcal{F}_T, \mathbb{P})$ and one looks for a pair (Y, Z) of $\mathbb{F}$-adapted processes verifying (7.24)–(7.25).

This form of (7.25), with a stochastic integral with respect to the forward process X rather than a Brownian motion, is the one which naturally appears in problems in stochastic control and mathematical finance (see, e.g., [39] for a discussion). The coefficients $b \colon \mathcal{W}_T \times \mathbb{R}^d \to \mathbb{R}^d$, $\sigma \colon \mathcal{W}_T \times \mathbb{R}^d \to \mathbb{R}^{d \times d}$, and $f \colon \mathcal{W}_T \times \mathbb{R}^d \times \mathbb{R} \times \mathbb{R}^d \to \mathbb{R}$ are assumed to verify the following standard assumptions [57]:

Assumption 7.7.1 (Assumptions on BSDE coefficients).   (i) For each $(x, y, z) \in \mathbb{R}^d \times \mathbb{R} \times \mathbb{R}^d$, $b(\cdot, \cdot, x)$, $\sigma(\cdot, \cdot, x)$ and $f(\cdot, \cdot, x, y, z)$ are non-anticipative functionals;

(ii) $b(t, \omega, \cdot)$, $\sigma(t, \omega, \cdot)$ and $f(t, \omega, \cdot, \cdot, \cdot)$ are Lipschitz continuous, uniformly with respect to $(t, \omega) \in \mathcal{W}_T$;

(iii) $E \int_0^T \left(|b(t, \cdot, 0)|^2 + |\sigma(t, \cdot, 0)|^2 + |f(t, \cdot, 0, 0, 0)|^2 \right) dt < \infty$;

(iv) $H \in L^2(\Omega, \mathcal{F}_T, \mathbb{P})$.

Under these assumptions, the forward SDE (7.24) has a unique strong solution X such that $E(\sup_{t \in [0,T]} |X(t)|^2) < \infty$ and $M(t) = \int_0^t \sigma(u, X_{u-}, X(u)) dW(u)$ is a square integrable martingale, with $M \in \mathcal{M}^2(W)$.

Theorem 7.7.1. *Assume*

$$\det \sigma(t, X_{t-}, X(t)) \neq 0 \quad dt \times d\mathbb{P}\text{-}a.e.$$

The FBSDE (7.24)–(7.25) *admits a unique solution* $(Y, Z) \in \mathcal{S}^{1,2}(M) \times \mathcal{L}^2(M)$ *such that* $E(\sup_{t \in [0,T]} |Y(t)|^2) < \infty$, *and* $Z(t) = \nabla_M Y(t)$.

Proof. Let us rewrite (7.25) in the classical form

$$Y(t) = H(X_T) - \int_t^T \underbrace{Z(u)\sigma(u, X_{u-}, X(u))}_{U(t)} \, dW(u)$$

$$- \int_t^T \underbrace{(f(u, X_{u-}, X(u), Y(u), Z(u)) + Z(u)b(u, X_{u-}, X(u)))}_{g(u, X_{u-}, X(u), Y(u), Z(u))} \, du. \quad (7.26)$$

Then, the map g also verifies Assumption 7.7.1 so, the classical result of Pardoux and Peng [57] implies the existence of a unique pair of $\mathbb{F}$-adapted processes (Y, U) such that

$$E\left(\sup_{t \in [0,T]} |Y(t)|^2 \right) + E\left(\int_0^T \|U(t)\|^2 dt \right) < \infty,$$

which verify (7.24)–(7.26). Using the decomposition (7.25), we observe that Y is an $\mathcal{F}_t^M$-semimartingale. Since $E(\sup_{t \in [0,T]} |Y(t)|^2) < \infty$, $Y \in \mathcal{S}^{1,2}(W)$ with

$\nabla_W Y(t) = Z(t)\sigma(t, X_{t-}, X(t))$. The non-singularity assumption on $\sigma(t, X_{t-}, X(t))$ implies that $\mathcal{F}_t^M = \mathcal{F}_t^W$. Let us define

$$Z(t) = U(t)\sigma(t, X_{t-}, X(t))^{-1}.$$

Then Z is $\mathcal{F}_t^M$-measurable, and

$$\|Z\|_{\mathcal{L}^2}^2(M) = E\left(\int_0^T Z dM\right)^2 = E\left(\int_0^T U dW\right)^2 = E\left(\int_0^T \|U(t)\|^2 dt\right) < \infty$$

so, $Z \in \mathcal{L}^2(M)$. Therefore, Y also belongs to $\mathcal{S}^{1,2}(M)$ and, by Proposition (7.6.1),

$$\nabla_W Y(t) = \nabla_M Y(t).\nabla_W M(t) = \nabla_M Y(t)\sigma(t, X_{t-}, X(t))$$

so,

$$\nabla_M Y(t) = \nabla_W Y(t)\sigma(t, X_{t-}, X(t))^{-1} = U(t)\sigma(t, X_{t-}, X(t))^{-1} = Z(t). \qquad \square$$

Chapter 8

Functional Kolmogorov equations

RAMA CONT and DAVID ANTOINE FOURNIÉ

One of the key topics in Stochastic Analysis is the deep link between Markov processes and partial differential equations, which can be used to characterize a diffusion process in terms of its infinitesimal generator [69]. Consider a second-order differential operator $L\colon C_b^{1,2}([0,T] \times \mathbb{R}^d) \to C_b^0([0,T] \times \mathbb{R}^d)$ defined, for test functions $f \in C^{1,2}([0,T] \times \mathbb{R}^d, \mathbb{R})$, by

$$Lf(t,x) = \frac{1}{2}\mathrm{tr}\big(\sigma(t,x) \cdot^t \sigma(t,x)\partial_x^2 f\big) + b(t,x)\partial_x f(t,x),$$

with continuous, bounded coefficients

$$b \in C_b^0([0,T] \times \mathbb{R}^d, \mathbb{R}^d), \quad \sigma \in C_b^0([0,T] \times \mathbb{R}^d, \mathbb{R}^{d \times n}).$$

Under various sets of assumptions on b and σ (such as Lipschitz continuity and linear growth; or uniform ellipticity, say, $\sigma^t\sigma(t,x) \geq \epsilon I_d$), the stochastic differential equation

$$dX(t) = b(t,X(t))dt + \sigma(t,X(t))dW(t)$$

has a unique weak solution $(\Omega, \mathcal{F}, (\mathcal{F}_t)_{t \geq 0}, W, \mathbb{P})$ and X is a Markov process under $\mathbb{P}$ whose evolution operator $(P_{t,s}^X, s \geq t \geq 0)$ has infinitesimal generator L. This weak solution $(\Omega, \mathcal{F}, (\mathcal{F}_t)_{t \geq 0}, W, \mathbb{P})$ is characterized by the property that, for every $f \in C_b^{1,2}([0,T] \times \mathbb{R}^d)$,

$$f(u,X(u)) - \int_0^u (\partial_t f + Lf)(t,X(t))dt$$

is a $\mathbb{P}$-martingale [69]. Indeed, applying Itô's formula yields

$$f(u,X(u)) = f(0,X(0)) + \int_0^u (\partial_t f + Lf)(t,X(t))dt$$

$$+ \int_0^u \partial_x f(t,X(t))\sigma(t,X(t))dW.$$

In particular, if $(\partial_t f + Lf)(t,x) = 0$ for every $t \in [0,T]$ and every $x \in \mathrm{supp}(X(t))$, then $M(t) = f(t,X(t))$ is a martingale.

More generally, for $f \in C^{1,2}([0,T] \times \mathbb{R}^d)$, $M(t) = f(t, X(t))$ is a local martingale if and only if f is a solution of

$$\partial_t f(t,x) + Lf(t,x) = 0 \quad \forall t \geq 0, \quad x \in \mathrm{supp}(X(t)).$$

This PDE is the Kolmogorov (backward) equation associated with X, whose solutions are the *space-time (L-)harmonic functions* associated with the differential operator L: f is space-time harmonic if and only if $f(t, X(t))$ is a (local) martingale.

The tools from Functional Itô Calculus may be used to extend these relations beyond the Markovian setting, to a large class of processes with path dependent characteristics. This leads us to a new class of partial differential equations on path space —*functional Kolmogorov equations*— which have many properties in common with their finite-dimensional counterparts and lead to new Feynman–Kac formulas for path dependent functionals of semimartingales. This is an exciting new topic, with many connections to other strands of stochastic analysis, and which is just starting to be explored [10, 14, 23, 28, 60].

8.1 Functional Kolmogorov equations and harmonic functionals

8.1.1 Stochastic differential equations with path dependent coefficients

Let $(\Omega = D([0,T], \mathbb{R}^d), \mathcal{F}^0)$ be the canonical space and consider a semimartingale X which can be represented as the solution of a stochastic differential equation whose coefficients are allowed to be path dependent, left continuous functionals:

$$dX(t) = b(t, X_t)dt + \sigma(t, X_t)dW(t), \tag{8.1}$$

where b and σ are non-anticipative functionals with values in $\mathbb{R}^d$ (resp., $\mathbb{R}^{d \times n}$) whose coordinates are in $\mathbb{C}^{0,0}_l(\Lambda_T)$ such that equation (8.1) has a unique weak solution $\mathbb{P}$ on $(\Omega, \mathcal{F}^0)$.

This class of processes is a natural 'path dependent' extension of the class of diffusion processes; various conditions (such as the functional Lipschitz property and boundedness) may be given for existence and uniqueness of solutions (see, e.g., [62]). We give here a sample of such a result, which will be useful in the examples.

Proposition 8.1.1 (Strong solutions for path dependent SDEs). *Assume that the non-anticipative functionals b and σ satisfy the following Lipschitz and linear growth conditions:*

$$\left| b(t,\omega) - b(t,\omega') \right| + \left| \sigma(t,\omega) - \sigma(t,\omega') \right| \leq K \sup_{s \leq t} \left| \omega(s) - \omega'(s) \right| \tag{8.2}$$

and

$$|b(t,\omega)| + |\sigma(t,\omega)| \le K\Big(1 + \sup_{s \le t} |\omega(s)| + |t|\Big), \tag{8.3}$$

for all $t \ge t_0$, $\omega, \omega' \in C^0([0,t], \mathbb{R}^d)$. Then, for any $\xi \in C^0([0,T], \mathbb{R}^d)$, the stochastic differential equation (8.1) has a unique strong solution X with initial condition $X_{t_0} = \xi_{t_0}$. The paths of X lie in $C^0([0,T], \mathbb{R}^d)$ and

(i) *there exists a constant C depending only on T and K such that, for $t \in [t_0, T]$,*

$$E\Big[\sup_{s \in [0,t]} |X(s)|^2\Big] \le C\Big(1 + \sup_{s \in [0,t_0]} |\xi(s)|^2\Big)e^{C(t-t_0)}; \tag{8.4}$$

(ii) $\int_0^{t-t_0} [|b(t_0 + s, X_{t_0+s})| + |\sigma(t_0 + s, X_{t_0+s})|^2] ds < +\infty$ *a.s.;*

(iii) $X(t) - X(t_0) = \int_0^{t-t_0} b(t_0 + s, X_{t_0+s}) ds + \sigma(t_0 + s, X_{t_0+s}) dW(s).$

We denote $\mathbb{P}_{(t_0,\xi)}$ the law of this solution.

Proofs of the uniqueness and existence, as well as the useful bound (8.4), can be found in [62].

Topological support of a stochastic process. In the statement of the link between a diffusion and the associated Kolmogorov equation, the domain of the PDE is related to the support of the random variable $X(t)$. In the same way, the support of the law (on the space of paths) of the process plays a key role in the path dependent extension of the Kolmogorov equation.

Recall that the *topological support* of a random variable X with values in a metric space is the smallest closed set supp(X) such that $\mathbb{P}(X \in \text{supp}(X)) = 1$.

Viewing a process X with continuous paths as a random variable on $(C^0([0,T], \mathbb{R}^d), \|\cdot\|_\infty)$ leads to the notion of *topological support* of a continuous process.

The support may be characterized by the following property: it is the set supp(X) of paths $\omega \in C^0([0,T], \mathbb{R}^d)$ for which every Borel neighborhood has strictly positive measure, i.e.,

$$\text{supp}(X) = \{\omega \in C^0([0,T], \mathbb{R}^d) \mid \forall \text{ Borel neigh. } V \text{ of } \omega, \ \mathbb{P}(X_T \in V) > 0\}.$$

Example 8.1.2. If W is a d-dimensional Wiener process with non-singular covariance matrix, then,

$$\text{supp}(W) = \{\omega \in C^0([0,T], \mathbb{R}^d), \omega(0) = 0\}.$$

Example 8.1.3. For an Itô process

$$X(t,\omega) = x + \int_0^t \sigma(t,\omega) dW,$$

if $\mathbb{P}(\sigma(t,\omega) \cdot^t \sigma(t,\omega) \ge \epsilon \, Id) = 1$, then supp$(X) = \{\omega \in C^0([0,T], \mathbb{R}^d), \omega(0) = x\}$.

A classical result due to Stroock–Varadhan [68] characterizes the topological support of a multi-dimensional diffusion process (for a simple proof see Millet and Sanz-Solé [54]).

Example 8.1.4 (Stroock–Varadhan support theorem). Let

$$b: [0,T] \times \mathbb{R}^d \longrightarrow \mathbb{R}^d, \quad \sigma: [0,T] \times \mathbb{R}^d \longrightarrow \mathbb{R}^{d \times d}$$

be Lipschitz continuous in x, uniformly on $[0,T]$. Then the law of the diffusion process

$$dX(t) = b(t, X(t))dt + \sigma(t, X(t))dW(t), \quad X(0) = x,$$

has a support given by the closure in $(C^0([0,T], \mathbb{R}^d), \|.\|_\infty)$ of the 'skeleton'

$$x + \{\omega \in H^1([0,T], \mathbb{R}^d), \exists h \in H^1([0,T], \mathbb{R}^d), \dot{\omega}(t) = b(t, \omega(t)) + \sigma(t, \omega(t))\dot{h}(t)\},$$

defined as the set of all solutions of the system of ODEs obtained when substituting $h \in H^1([0,T], \mathbb{R}^d)$ for W in the SDE.

We will denote by $\Lambda(T, X)$ the set of all stopped paths obtained from paths in $\mathrm{supp}(X)$:

$$\Lambda(T, X) = \{(t, \omega) \in \Lambda_T, \omega \in \mathrm{supp}(X)\}. \tag{8.5}$$

For a continuous process X, $\Lambda(T, X) \subset \mathcal{W}_T$.

8.1.2   Local martingales and harmonic functionals

Functionals of a process X which have the local martingale property play an important role in stochastic control theory [17] and mathematical finance. By analogy with the Markovian case, we call such functionals $\mathbb{P}$-harmonic functionals:

Definition 8.1.5 ($\mathbb{P}$-harmonic functionals). *A non-anticipative functional F is called $\mathbb{P}$-harmonic if $Y(t) = F(t, X_t)$ is a $\mathbb{P}$-local martingale.*

The following result characterizes smooth $\mathbb{P}$-harmonic functionals as solutions to a *functional Kolmogorov equation*, see [10].

Theorem 8.1.6 (Functional equation for $\mathcal{C}^{1,2}$ martingales). *If $F \in \mathbb{C}_b^{1,2}(\mathcal{W}_T)$ and $\mathcal{D}F \in \mathbb{C}_l^{0,0}$, then $Y(t) = F(t, X_t)$ is a local martingale if and only if F satisfies*

$$\mathcal{D}F(t, \omega_t) + b(t, \omega_t)\nabla_\omega F(t, \omega_t) + \frac{1}{2}\mathrm{tr}\left[\nabla_\omega^2 F(t, \omega)\sigma^t\sigma(t, \omega)\right] = 0 \tag{8.6}$$

on the topological support of X in $(C^0([0,T], \mathbb{R}^d), \|.\|_\infty)$.

Proof. If $F \in \mathbb{C}_b^{1,2}(\mathcal{W}_T)$ is a solution of (8.6), the functional Itô formula (Theorem 6.2.3) applied to $Y(t) = F(t, X_t)$ shows that Y has the semimartingale decomposition

$$Y(t) = Y(0) + \int_0^t Z(u)du + \int_0^t \nabla_\omega F(u, X_u)\sigma(u, X_u)dW(u),$$

where

$$Z(t) = \mathcal{D}F(t, X_t) + b(t, X_t)\nabla_\omega F(t, X_t) + \frac{1}{2}\text{tr}\big(\nabla_\omega^2 F(t, X_t)\sigma^t\sigma(t, X_t)\big).$$

Equation (8.6) then implies that the finite variation term in (6.2) is almost surely zero. So, $Y(t) = Y(0) + \int_0^t \nabla_\omega F(u, X_u).\sigma(u, X_u)dW(u)$ and thus Y is a continuous local martingale.

Conversely, assume that $Y = F(X)$ is a local martingale. Let $A(t, \omega) = \sigma(t, \omega)^t\sigma(t, \omega)$. Then, by Lemma 5.1.5, Y is left-continuous. Suppose (8.6) is not satisfied for some $\omega \in \text{supp}(X) \subset C^0([0, T], \mathbb{R}^d)$ and some $t_0 < T$. Then, there exist $\eta > 0$ and $\epsilon > 0$ such that

$$\left|\mathcal{D}F(t, \omega) + b(t, \omega)\nabla_\omega F(t, \omega) + \frac{1}{2}\text{tr}\big[\nabla_\omega^2 F(t, \omega)A(t, \omega)\big]\right| > \epsilon,$$

for $t \in [t_0 - \eta, t_0]$, by left-continuity of the expression. Also, there exist a neighborhood V of ω in $C^0([0, T], \mathbb{R}^d), \|.\|_\infty)$ such that, for all $\omega' \in V$ and $t \in [t_0 - \eta, t_0]$,

$$\left|\mathcal{D}F(t, \omega') + b(t, \omega')\nabla_\omega F(t, \omega') + \frac{1}{2}\text{tr}\big[\nabla_\omega^2 F(t, \omega')A(t, \omega')\big]\right| > \frac{\epsilon}{2}.$$

Since $\omega \in \text{supp}(X)$, $\mathbb{P}(X \in V) > 0$, and so

$$\left\{(t, \omega) \in \mathcal{W}_T, \left|\mathcal{D}F(t, \omega) + b(t, \omega)\nabla_\omega F(t, \omega) + \frac{1}{2}\text{tr}\big(\nabla_\omega^2 F(t, \omega)A(t, \omega)\big)\right| > \frac{\epsilon}{2}\right\}$$

has non-zero measure with respect to $dt \times d\mathbb{P}$. Applying the functional Itô formula (6.2) to the process $Y(t) = F(t, X_t)$ then leads to a contradiction because, as a continuous local martingale, its finite variation component should be zero $dt \times d\mathbb{P}$-a.e. $\qquad\square$

In the case where $F(t, \omega) = f(t, \omega(t))$ and the coefficients b and σ are not path dependent, this equation reduces to the well-known backward Kolmogorov equation [46].

Remark 8.1.7 (Relation with infinite-dimensional Kolmogorov equations in Hilbert spaces). Note that the second-order operator ∇_ω^2 is an iterated directional derivative, and not a second-order Fréchet or "H-derivative" so, this equation is different from the infinite-dimensional Kolmogorov equations in Hilbert spaces as described, for example, by Da Prato and Zabczyk [16]. The relation between these two classes of infinite-dimensional Kolmogorov equations has been studied recently by Flandoli and Zanco [28], who show that, although one can partly recover some results for (8.6) using the theory of infinite-dimensional Kolmogorov equations in Hilbert spaces, this can only be done at the price of higher regularity requirements (not the least of them being Fréchet differentiability), which exclude many interesting examples.

When $X = W$ is a d-dimensional Wiener process, Theorem 8.1.6 gives a characterization of regular Brownian local martingales as solutions of a 'functional heat equation'.

Corollary 8.1.8 (Harmonic functionals on $\mathcal{W}_T$). *Let $F \in \mathbb{C}_b^{1,2}(\mathcal{W}_T)$. Then, $(Y(t) = F(t, W_t), t \in [0, T])$ is a local martingale if and only if for every $t \in [0, T]$ and every $\omega \in C^0([0, T], \mathbb{R}^d)$,*

$$\mathcal{D}F(t, \omega) + \frac{1}{2}\,\mathrm{tr}\big(\nabla_\omega^2 F(t, \omega)\big) = 0.$$

8.1.3   Sub-solutions and super-solutions

By analogy with the case of finite-dimensional parabolic PDEs, we introduce the notion of sub- and super-solution to (8.6).

Definition 8.1.9 (Sub-solutions and super-solutions). *$F \in \mathbb{C}^{1,2}(\Lambda_T)$ is called a super-solution of (8.6) on a domain $U \subset \Lambda_T$ if, for all $(t, \omega) \in U$,*

$$\mathcal{D}F(t, \omega) + b(t, \omega) \cdot \nabla_\omega F(t, \omega) + \frac{1}{2}\mathrm{tr}\big[\nabla_\omega^2 F(t, \omega)\sigma^{\mathrm{t}}\sigma(t, \omega)\big] \leq 0. \qquad (8.7)$$

Similarly, $F \in \mathbb{C}_{\mathrm{loc}}^{1,2}(\Lambda_T)$ is called a sub-solution of (8.6) on U if, for all $(t, \omega) \in U$,

$$\mathcal{D}F(t, \omega) + b(t, \omega) \cdot \nabla_\omega F(t, \omega) + \frac{1}{2}\mathrm{tr}\big[\nabla_\omega^2 F(t, \omega)\sigma^{\mathrm{t}}\sigma(t, \omega)\big] \geq 0. \qquad (8.8)$$

The following result shows that $\mathbb{P}$-integrable sub- and super-solutions have a natural connection with $\mathbb{P}$-submartingales and $\mathbb{P}$-supermartingales.

Proposition 8.1.10. *Let $F \in \mathbb{C}_b^{1,2}(\Lambda_T)$ be such that, for all $t \in [0, T]$, we have $E(|F(t, X_t)|) < \infty$. Set $Y(t) = F(t, X_t)$.*

(i) *Y is a submartingale if and only if F is a sub-solution of (8.6) on $\mathrm{supp}(X)$.*

(ii) *Y is a supermartingale if and only if F is a super-solution of (8.6) on $\mathrm{supp}(X)$.*

Proof. If $F \in \mathbb{C}^{1,2}$ is a super-solution of (8.6), the application of the Functional Itô formula (Theorem 6.2.3, Eq. (6.2)) gives

$$Y(t) - Y(0) = \int_0^t \nabla_\omega F(u, X_u)\sigma(u, X_u)dW(u)$$

$$+ \int_0^t \left(\mathcal{D}F(u, X_u) + \nabla_\omega F(u, X_u)b(u, X_u) + \frac{1}{2}\mathrm{tr}\big(\nabla_\omega^2 F(u, X_u)A(u, X_u)\big) \right) du \text{ a.s.,}$$

where the first term is a local martingale and, since F is a super-solution, the finite variation term (second line) is a decreasing process, so Y is a supermartingale.

Conversely, let $F \in \mathbb{C}^{1,2}(\Lambda_T)$ be such that $Y(t) = F(t, X_t)$ is a supermartingale. Let $Y = M - H$ be its Doob–Meyer decomposition, where M is a local martingale and H an increasing process. By the functional Itô formula, Y has the above decomposition so it is a continuous supermartingale. By the uniqueness of the Doob–Meyer decomposition, the second term is thus a decreasing process so,

$$\mathcal{D}F(t, X_t) + \nabla_\omega F(t, X_t)b(t, X_t) + \frac{1}{2}\mathrm{tr}(\nabla_\omega^2 F(t, X_t)A(t, X_t)) \leq 0 \; dt \times d\mathbb{P}\text{-a.e.}$$

This implies that the set

$$S = \{\omega \in C^0([0, T], \mathbb{R}^d) \mid (8.7) \text{ holds for all } t \in [0, T]\}$$

includes any Borel set $B \subset C^0([0, T], \mathbb{R}^d)$ with $\mathbb{P}(X_T \in B) > 0$. Using continuity of paths,

$$S = \bigcap_{t \in [0,T] \cap \mathbb{Q}} \{\omega \in C^0([0, T], \mathbb{R}^d) \mid (8.7) \text{ holds for } (t, \omega)\}.$$

Since $F \in \mathbb{C}^{1,2}(\Lambda_T)$, this is a closed set in $(C^0([0, T], \mathbb{R}^d), \|\cdot\|_\infty)$ so, S contains the topological support of X, i.e., (8.7) holds on $\Lambda_T(X)$. □

8.1.4 Comparison principle and uniqueness

A key property of the functional Kolmogorov equation (8.6) is the comparison principle. As in the finite-dimensional case, this requires imposing an integrability condition.

Theorem 8.1.11 (Comparison principle). *Let $\underline{F} \in \mathbb{C}^{1,2}(\Lambda_T)$ be a sub-solution of (8.6) and $\overline{F} \in \mathbb{C}^{1,2}(\Lambda_T)$ be a super-solution of (8.6) such that, for every $\omega \in C^0([0, T], \mathbb{R}^d)$, $\underline{F}(T, \omega) \leq \overline{F}(T, \omega)$, and*

$$E\left(\sup_{t \in [0,T]} |\underline{F}(t, X_t) - \overline{F}(t, X_t)|\right) < \infty.$$

Then, $\forall t \in [0, T], \forall \omega \in \mathrm{supp}(X)$,

$$\underline{F}(t, \omega) \leq \overline{F}(t, \omega). \tag{8.9}$$

Proof. Since $F = \underline{F} - \overline{F} \in \mathbb{C}^{1,2}(\Lambda_T)$ is a sub-solution of (8.6), by Proposition 8.1.10, the process Y defined by $Y(t) = F(t, X_t)$ is a submartingale. Since $Y(T) = \underline{F}(T, X_T) - \overline{F}(T, X_T) \leq 0$, we have

$$\forall t \in [0, T[, \quad Y(t) \leq 0 \quad \mathbb{P}\text{-a.s.}$$

Define

$$S = \{\omega \in C^0([0, T], \mathbb{R}^d), \forall t \in [0, T], \underline{F}(t, \omega) \leq \overline{F}(t, \omega)\}.$$

Using continuity of paths, and since $F \in \mathbb{C}^{0,0}(W_T)$,

$$S = \bigcap_{t \in \mathbb{Q} \cap [0,T]} \{\omega \in \mathrm{supp}(X), \underline{F}(t, \omega) \leq \overline{F}(t, \omega)\}$$

is closed in $(C^0([0,T], \mathbb{R}^d), \|.\|_\infty)$. If (8.9) does not hold, then there exist $t_0 \in [0, T[$ and $\omega \in \mathrm{supp}(X)$ such that $F(t_0, \omega) > 0$. Since $O = \mathrm{supp}(X) - S$ is a non-empty open subset of $\mathrm{supp}(X)$, there exists an open set $A \subset O \subset \mathrm{supp}(X)$ containing ω and $h > 0$ such that, for every $t \in [t_0 - h, t]$ and $\omega \in A$, $F(t, \omega) > 0$. But $\mathbb{P}(X \in A) > 0$ implies that $\int dt \times d\mathbb{P} \, 1_{F(t,\omega) > 0} > 0$, which contradicts the above assumption. □

The following uniqueness result for $\mathbb{P}$-uniformly integrable solutions of the Functional Kolmogorov equation (8.6) is a straightforward consequence of the comparison principle.

Theorem 8.1.12 (Uniqueness of solutions). *Let $H : (C^0([0,T], \mathbb{R}^d), \| \cdot \|_\infty) \to \mathbb{R}$ be a continuous functional, and let $F^1, F^2 \in \mathbb{C}_b^{1,2}(\mathcal{W}_T)$ be solutions of the Functional Kolmogorov equation* (8.6) *verifying*

$$F^1(T, \omega) = F^2(T, \omega) = H(\omega) \tag{8.10}$$

and

$$E\left[\sup_{t \in [0,T]} |F^1(t, X_t) - F^2(t, X_t)| \right] < \infty. \tag{8.11}$$

Then, they coincide on the topological support of X, i.e.,

$$\forall \omega \in \mathrm{supp}(X), \quad \forall t \in [0,T], \quad F^1(t, \omega) = F^2(t, \omega).$$

The integrability condition in this result (or some variation of it) is required for uniqueness: indeed, even in the finite-dimensional setting of a heat equation in $\mathbb{R}^d$, one cannot do without an integrability condition with respect to the heat kernel. Without this condition, we can only assert that $F(t, X_t)$ is a local martingale, so it is not uniquely determined by its terminal value and counterexamples to uniqueness are easy to construct.

Note also that uniqueness holds on the support of X, the process associated with the (path dependent) operator appearing in the Kolmogorov equation. The local martingale property of $F(X)$ implies no restriction on the behavior of F outside $\mathrm{supp}(X)$ so, there is no reason to expect uniqueness or comparison of solutions on the whole path space.

8.1.5   Feynman–Kac formula for path dependent functionals

As observed above, any $\mathbb{P}$-uniformly integrable solution F of the functional Kolmogorov equation yields a $\mathbb{P}$-martingale $Y(t) = F(t, X_t)$. Combined with the uniqueness result (Theorem 8.1.12), this leads to an extension of the well-known Feynman–Kac representation to the path dependent case.

Theorem 8.1.13 (A Feynman–Kac formula for path dependent functionals). *Let $H : (C^0([0,T]), \| \cdot \|_\infty) \to \mathbb{R}$ be continuous. If, for every $\omega \in C^0([0,T])$, $F \in \mathbb{C}_b^{1,2}(\mathcal{W}_T)$ verifies*

$$\mathcal{D}F(t, \omega) + b(t, \omega) \cdot \nabla_\omega F(t, \omega) + \frac{1}{2}\mathrm{tr}\left[\nabla_\omega^2 F(t, \omega)\sigma^{\mathrm{t}}\sigma(t, \omega)\right] = 0,$$

$$F(T, \omega) = H(\omega), \quad E\left[\sup_{t \in [0,T]} |F(t, X_t)|\right] < \infty,$$

then F has the probabilistic representation

$$\forall \omega \in \mathrm{supp}(X), \quad F(t, \omega) = E^{\mathbb{P}}\big[H(X_T) \,|\, X_t = \omega_t\big] = E^{\mathbb{P}_{(t,\omega)}}[H(X_T)],$$

where $\mathbb{P}_{(t,\omega)}$ is the law of the unique solution of the SDE $dX(t) = b(t, X_t)dt + \sigma(t, X_t)dW(t)$ *with initial condition $X_t = \omega_t$. In particular,*

$$F(t, X_t) = E^{\mathbb{P}}\big[H(X_T) \,|\, \mathcal{F}_t\big] \quad dt \times d\mathbb{P}\text{-}a.s.$$

8.2 FBSDEs and semilinear functional PDEs

The relation between stochastic processes and partial differential equations extends to the semilinear case: in the Markovian setting, there is a well-known relation between semilinear PDEs of parabolic type and Markovian forward-backward stochastic differential equations (FBSDE), see [24, 58, 74]. The tools of Functional Itô Calculus allows to extend the relation between semilinear PDEs and FBSDEs beyond the Markovian case, to non-Markovian FBSDEs with path dependent coefficients.

Consider the forward-backward stochastic differential equation with path dependent coefficients

$$X(t) = X(0) + \int_0^t b\big(u, X_{u-}, X(u)\big)du + \int_0^t \sigma\big(u, X_{u-}, X(u)\big)dW(u), \quad (8.12)$$

$$Y(t) = H(X_T) + \int_t^T f\big(s, X_{s-}, X(s), Y(s), Z(s)\big)ds - \int_t^T Z dM, \quad (8.13)$$

whose coefficients

$$b \colon \mathcal{W}_T \times \mathbb{R}^d \longrightarrow \mathbb{R}^d, \quad \sigma \colon \mathcal{W}_T \times \mathbb{R}^d \longrightarrow \mathbb{R}^{d \times d}, \quad f \colon \mathcal{W}_T \times \mathbb{R}^d \times \mathbb{R} \times \mathbb{R}^d \longrightarrow \mathbb{R}^d$$

are assumed to verify Assumption 7.7.1. Under these assumptions,

(i) the forward SDE given by (8.12) has a unique strong solution X with $E(\sup_{t \in [0,T]} |X(t)|^2) < \infty$, whose martingale part we denote by M, and

(ii) the FBSDE (8.12)–(8.13) admits a unique solution $(Y, Z) \in \mathcal{S}^{1,2}(M) \times \mathcal{L}^2(M)$ with $E(\sup_{t \in [0,T]} |Y(t)|^2) < \infty$.

Define

$$B(t, \omega) = b\big(t, \omega_{t-}, \omega(t)\big), \quad A(t, \omega) = \sigma\big(t, \omega_{t-}, \omega(t)\big)^{\mathsf{t}} \sigma\big(t, \omega_{t-}, \omega(t)\big).$$

In the 'Markovian' case, where the coefficients are not path dependent, the solution of the FBSDE can be obtained by solving a semilinear PDE: if $f \in C^{1,2}([0,T] \times \mathbb{R}^d])$ is a solution of

$$\partial_t f(t,\omega) + f\big(t,x,f(t,x),\nabla f(t,x)\big) + b(t,x)\nabla f(t,x) + \frac{1}{2}\mathrm{tr}\big[a(t,\omega)\cdot\nabla^2 f(t,x)\big] = 0,$$

then a solution of the FBSDE is given by $(Y(t),Z(t)) = (f(t,X(t)),\nabla f(t,X(t)))$. Here, the function u 'decouples' the solution of the backward equation from the solution of the forward equation. In the general path dependent case (8.12)–(8.13) we have a similar object, named the 'decoupling random field' in [51] following earlier ideas from [50].

   The following result shows that such a 'decoupling random field' for the path dependent FBSDE (8.12)–(8.13) may be constructed as a solution of a semilinear functional PDE, analogously to the Markovian case.

Theorem 8.2.1 (FBSDEs as semilinear path dependent PDEs). *Let $F \in \mathbb{C}^{1,2}_{\mathrm{loc}}(\mathcal{W}_T)$ be a solution of the path dependent semilinear equation*

$$\mathcal{D}F(t,\omega) + f\big(t,\omega_{t-},\omega(t),F(t,\omega),\nabla_\omega F(t,\omega)\big)$$

$$+ B(t,\omega)\cdot\nabla_\omega F(t,\omega) + \frac{1}{2}\mathrm{tr}\big[A(t,\omega).\nabla^2_\omega F(t,\omega)\big] = 0,$$

for $\omega \in \mathrm{supp}(X)$, $t \in [0,T]$, and with $F(T,\omega) = H(\omega)$. Then, the pair of processes (Y,Z) given by

$$\big(Y(t),Z(t)\big) = \big(F(t,X_t),\nabla_\omega F(t,X_t)\big)$$

is a solution of the FBSDE (8.12)–(8.13).

Remark 8.2.2. In particular, under the conditions of Theorem 8.2.1 the random field $u\colon [0,T] \times \mathbb{R}^d \times \Omega \to \mathbb{R}$ defined by

$$u(t,x,\omega) = F(t,X_{t-}(t,\omega)) + x1_{[t,T]})$$

is a 'decoupling random field' in the sense of [51].

Proof of Theorem 8.2.1. Let $F \in \mathbb{C}^{1,2}_{\mathrm{loc}}(\mathcal{W}_T)$ be a solution of the semilinear equation above. Then the processes (Y,Z) defined above are $\mathbb{F}$-adapted. Applying the functional Itô formula to $F(t,X_t)$ yields

$$H(X_T) - F(t,X_t) = \int_t^T \nabla_\omega F(u,X_u)dM$$

$$+ \int_t^T \left(\mathcal{D}F(u,X_u) + B(u,X_u)\nabla_\omega F(u,X_u) + \frac{1}{2}\mathrm{tr}\big[A(u,X_u)\nabla^2_\omega F(u,X_u)\big]\right)du.$$

Since F is a solution of the equation, the term in the last integral is equal to

$$-f(t,X_{t-},X(t),F(t,X_t),\nabla_\omega F(t,X_t)) = -f(t,X_{t-},X(t),Y(t),Z(t)).$$

So, $(Y(t), Z(t)) = (F(t, X_t), \nabla_\omega F(t, X_t))$ verifies

$$Y(t) = H(X_T) - \int_t^T f\big(s, X_{s-}, X(s), Y(s), Z(s)\big)ds - \int_t^T Z dM.$$

Therefore, (Y, Z) is a solution to the FBSDE (8.12)–(8.13). $\qquad\square$

This result provides the "Hamiltonian" version of the FBSDE (8.12)–(8.13), in the terminology of [74].

8.3 Non-Markovian stochastic control and path dependent HJB equations

An interesting application of the Functional Itô Calculus is the study of non-Markovian stochastic control problems, where both the evolution of the state of the system and the control policy are allowed to be path dependent.

Non-Markovian stochastic control problems were studied by Peng [59], who showed that, when the characteristics of the controlled process are allowed to be stochastic (i.e., path dependent), the optimality conditions can be expressed in terms of a stochastic Hamilton–Jacobi–Bellman equation. The relation with FBSDEs is explored systematically in the monograph [74].

We formulate a mathematical framework for stochastic optimal control where the evolution of the state variable, the control policy and the objective are allowed to be non-anticipative path dependent functionals, and show how the Functional Itô formula in Theorem 6.2.3 characterizes the value functional as the solution to a *functional Hamilton–Jacobi–Bellman equation.*

This section is based on [32].

Let W be a d-dimensional Brownian motion on a probability space $(\Omega, \mathcal{F}, \mathbb{P})$ and $\mathbb{F} = (\mathcal{F}_t)_{t\geq 0}$ the $\mathbb{P}$-augmented natural filtration of W.

Let C be a subset of $\mathbb{R}^m$. We consider the controlled stochastic differential equation

$$dX(t) = b\big(t, X_t, \alpha(t)\big)dt + \sigma\big(t, X_t, \alpha(t)\big)dW(t), \tag{8.14}$$

where the coefficients

$$b: \mathcal{W}_T \times C \longrightarrow \mathbb{R}^d, \quad \sigma: \mathcal{W}_T \times C \longrightarrow \mathbb{R}^{d\times d}$$

are allowed to be path dependent and assumed to verify the conditions of Proposition 8.1.1 uniformly with respect to $\alpha \in C$, and the control α belongs to the set $\mathcal{A}(\mathbb{F})$ of $\mathbb{F}$-progressively measurable processes with values in C verifying

$$E\left(\int_0^T \|\alpha(t)\|^2 dt\right) < \infty.$$

Then, for any initial condition $(t, \omega) \in \mathcal{W}_T$ and for each $\alpha \in \mathcal{A}(\mathbb{F})$, the stochastic differential equation (8.14) admits a unique strong solution, which we denote by $X^{(t,\omega,\alpha)}$.

Let $g \colon (C^0([0, T], \mathbb{R}^d), \|.\|_\infty) \to \mathbb{R}$ be a continuous map representing a terminal cost and $L \colon \mathcal{W}_T \times C \to L(t, \omega, u)$ a non-anticipative functional representing a possibly path dependent running cost. We assume

(i) $\exists K > 0$ such that $-K \leq g(\omega) \leq K(1 + \sup_{s \in [0,t]} |\omega(s)|^2)$;

(ii) $\exists K' > 0$ such that $-K' \leq L(t, \omega, u) \leq K'(1 + \sup_{s \in [0,t]} |\omega(s)|^2)$.

Then, thanks to (8.4), the cost functional

$$J(t, \omega, \alpha) = E\left[g(X_T^{(t,\omega,\alpha)}) + \int_t^T L(s, X_s^{(t,\omega,\alpha)}, \alpha(s)) ds\right] < \infty$$

for $(t, \omega, \alpha) \in \mathcal{W}_T \times \mathcal{A}$ defines a non-anticipative map $J \colon \mathcal{W}_T \times \mathcal{A}(\mathbb{F}) \to \mathbb{R}$.

We consider the optimal control problem

$$\inf_{\alpha \in \mathcal{A}(\mathbb{F})} J(t, \omega, \alpha)$$

whose value functional is denoted, for $(t, \omega) \in \mathcal{W}_T$,

$$V(t, \omega) = \inf_{\alpha \in \mathcal{A}(\mathbb{F})} J(t, \omega, \alpha). \tag{8.15}$$

From the above assumptions and the estimate (8.4), $V(t, \omega)$ verifies

$$\exists K'' > 0 \text{ such that } -K'' \leq V(t, \omega) \leq K''\left(1 + \sup_{s \in [0,t]} |\omega(s)|^2\right).$$

Introduce now the *Hamiltonian* associated to the control problem [74] as the non-anticipative functional $H \colon \mathcal{W}_T \times \mathbb{R}^d \times S_d^+ \to \mathbb{R}$ given by

$$H(t, \omega, \rho, A) = \inf_{u \in C} \left\{\frac{1}{2} \mathrm{tr}\left(A \sigma(t, \omega, u)^t \sigma(t, \omega, u)\right) + b(t, \omega, u)\rho + L(t, \omega, u)\right\}.$$

It is readily observed that, for any $\alpha \in \mathcal{A}(\mathbb{F})$, the process

$$V\left(t, X_t^{(t_0,\omega,\alpha)}\right) + \int_{t_0}^t L\left(s, X_s^{(t_0,\omega,\alpha)}, \alpha(s)\right) ds$$

has the submartingale property. The martingale optimality principle [17] then characterizes an optimal control α^* as one for which this process is a martingale.

We can now use the Functional Itô Formula (Theorem 6.2.3) to give a sufficient condition for a functional U to be equal to the value functional V and for a control α^* to be optimal. This condition takes the form of a path dependent Hamilton–Jacobi–Bellman equation.

Theorem 8.3.1 (Verification Theorem). *Let $U \in \mathbb{C}_b^{1,2}(\mathcal{W}_T)$ be a solution of the functional HJB equation,*

$$\forall (t,\omega) \in \mathcal{W}_T, \quad \mathcal{D}U(t,\omega) + H(t,\omega, \nabla_\omega U(t,\omega), \nabla_\omega^2 U(t,\omega)) = 0, \qquad (8.16)$$

satisfying $U(T,\omega) = g(\omega)$ and the quadratic growth condition

$$|U(t,\omega)| \leq C \sup_{s \leq t} |\omega(s)|^2. \qquad (8.17)$$

Then, for any $\omega \in C^0([0,T], \mathbb{R}^d)$ and any admissible control $\alpha \in \mathcal{A}(\mathbb{F})$,

$$U(t,\omega) \leq J(t,\omega,\alpha). \qquad (8.18)$$

If, furthermore, for $\omega \in C^0([0,T], \mathbb{R}^d)$ there exists $\alpha^* \in \mathcal{A}(\mathbb{F})$ such that

$$H\big(s, X_s^{(t,\omega,\alpha^*)}, \nabla_\omega U\big(s, X_s^{(t,\omega,\alpha^*)}\big), \nabla_\omega^2 U\big(s, X_s^{(t,\omega,\alpha^*)}\big)\big) \qquad (8.19)$$

$$= \frac{1}{2} \operatorname{tr}\big[\nabla_\omega^2 U\big(s, X_s^{(t,\omega,\alpha^*)}\big) \sigma\big(s, X_s^{(t,\omega,\alpha^*)}, \alpha^*(s)\big)^{\mathsf{t}} \sigma\big(s, X_s^{(t,\omega,\alpha^*)}, \alpha^*(s)\big)\big]$$

$$+ \nabla_\omega U\big(s, X_s^{(t,\omega,\alpha^*)}\big) b\big(s, X_s^{(t,\omega,\alpha^*)}, \alpha^*(s)\big) + L\big(s, X_s^{(t,\omega,\alpha^*)}, \alpha^*(s)\big)$$

$ds \times d\mathbb{P}$-a.e. for $s \in [t,T]$, then $U(t,\omega) = V(t,\omega) = J(t,\omega,\alpha^*)$ and α^* is an optimal control.

Proof. Let $\alpha \in \mathcal{A}(\mathbb{F})$ be an admissible control, $t < T$ and $\omega \in C^0([0,T], \mathbb{R}^d)$. Applying the Functional Itô Formula yields, for $s \in [t,T]$,

$$U\big(s, X_s^{(t,\omega,\alpha)}\big) - U(t,\omega)$$

$$= \int_t^s \nabla_\omega U\big(u, X_u^{(t,\omega,\alpha)}\big) \sigma\big(u, X_u^{(t,\omega,\alpha)}, \alpha(u)\big) dW(u)$$

$$+ \int_t^s \mathcal{D}U\big(u, X_u^{(t,\omega,\alpha)}\big) + \nabla_\omega U\big(u, X_u^{(t,\omega,\alpha)}\big) b\big(u, X_u^{(t,\omega,\alpha)}, \alpha(u)\big) du$$

$$+ \int_t^s \frac{1}{2} \operatorname{tr}\Big(\nabla_\omega^2 U\big(u, X_u^{(t,\omega,\alpha)}\big) \sigma^{\mathsf{t}} \sigma\big(u, X_u^{(t,\omega,\alpha)}\big)\Big) du.$$

Since U verifies the functional Hamilton–Jacobi–Bellman equation, we have

$$U\big(s, X_s^{(t,\omega,\alpha)}\big) - U(t,\omega) \geq \int_t^s \nabla_\omega U\big(u, X_u^{(t,\omega,\alpha)}\big) \sigma\big(u, X_u^{(t,\omega,\alpha)}, \alpha(u)\big) dW(u)$$

$$- \int_t^s L\big(u, X_u^{(t,\omega,\alpha)}, \alpha(u)\big) du.$$

In other words, $U(s, X_s^{(t,\omega,\alpha)}) - U(t,\omega) + \int_t^s L(u, X_u^{(t,\omega,\alpha)}, \alpha(u)) du$ is a local submartingale. The estimate (8.17) and the L^2 estimate (8.4) guarantee that it is actually a submartingale, hence, taking $s \to T$, the left continuity of U yields

$$E\Big[g\big(X_T^{(t,\omega,\alpha)}\big) + \int_0^{T-t} L\big(t+u, X_{t+u}^{(t,\omega,\alpha)}, \alpha(u)\big) du\Big] \geq U(t,\omega).$$

This being true for any $\alpha \in \mathcal{A}(\mathbb{F})$, we conclude that $U(t, \omega) \leq J(t, \omega, \alpha)$.

Assume now that $\alpha^* \in \mathcal{A}(\mathbb{F})$ verifies (8.19). Taking $\alpha = \alpha^*$ transforms inequalities to equalities, submartingale to martingale, and hence establishes the second part of the theorem. $\qquad\qquad\qquad\qquad\qquad\qquad\qquad\qquad\qquad\qquad\qquad\square$

This proof can be adapted [32] to the case where all coefficients, except σ, depend on the quadratic variation of X as well, using the approach outlined in Section 6.3. The subtle point is that if σ depends on the control, the Functional Itô Formula (Theorem 6.2.3) does not apply to $U(s, X_s^\alpha, [X^\alpha]_s)$ because $d[X^\alpha](s)/ds$ would not necessarily admit a right continuous representative.

8.4   Weak solutions

The above results are 'verification theorems' which, assuming the existence of a solution to the Functional Kolmogorov Equation with a given regularity, derive a probabilistic property or probabilistic representation of this solution. The key issue when applying such results is to be able to prove the regularity conditions needed to apply the Functional Itô Formula. In the case of linear Kolmogorov equations, this is equivalent to constructing, given a functional H, a *smooth* version of the conditional expectation $E[H|\mathcal{F}_t]$, i.e., a functional representation admitting vertical and horizontal derivatives.

As with the case of finite-dimensional PDEs, such *strong solutions* —with the required differentiability— may fail to exist in many examples of interest and, even if they exist, proving pathwise regularity is not easy in general and requires imposing many smoothness conditions on the terminal functional H, due to the absence of any regularizing effect as in the finite-dimensional parabolic case [13, 61].

A more general approach is provided by the notion of *weak solution*, which we now define, using the tools developed in Chapter 7. Consider, as above, a semimartingale X defined as the solution of a stochastic differential equation

$$X(t) = X(0) + \int_0^t b(u, X_u)du + \int_0^t \sigma(u, X_u)dW(u) = X(0) + \int_0^t b(u, X_u)du + M(t)$$

whose coefficients b, σ are non-anticipative path dependent functionals, verifying the assumptions of Proposition 8.1.1. We assume that M, the martingale part of X, is a square integrable martingale.

We can then use the notion of weak derivative ∇_M on the Sobolev space $\mathcal{S}^{1,2}(M)$ of semimartingales introduced in Section 7.5.

The idea is to introduce a probability measure $\mathbb{P}$ on $D([0, T], \mathbb{R}^d)$, under which X is a semimartingale, and requiring (8.6) to hold $dt \times d\mathbb{P}$-a.e.

For a classical solution $F \in \mathbb{C}_b^{1,2}(\mathcal{W}_T)$, requiring (8.6) to hold on the support of M is equivalent to requiring

$$\mathcal{A}F(t, X_t) = \mathcal{D}F(t, X_t) + \nabla_\omega F(t, X_t)b(t, X_t) + \frac{1}{2}\text{tr}\big[\nabla^2_\omega F(t, X_t)\sigma^t\sigma(t, X_t)\big] = 0,$$

$$\tag{8.20}$$

$dt \times d\mathbb{P}$-a.e.

Let $\mathcal{L}^2(\mathbb{F}, dt \times d\mathbb{P})$ be the space of $\mathbb{F}$-adapted processes $\phi\colon [0, T] \times \Omega \to \mathbb{R}$ with

$$E\left(\int_0^T |\phi(t)|^2 dt\right) < \infty,$$

and $\mathcal{A}^2(\mathbb{F})$ the space of absolutely continuous processes whose Radon–Nikodým derivative lies in $\mathcal{L}^2(\mathbb{F}, dt \times d\mathbb{P})$:

$$\mathcal{A}^2(\mathbb{F}) = \left\{\Phi \in \mathcal{L}^2(\mathbb{F}, dt \times d\mathbb{P}) \mid \frac{d\Phi}{dt} \in \mathcal{L}^2(\mathbb{F}, dt \times d\mathbb{P})\right\}.$$

Using the density of $\mathcal{A}^2(\mathbb{F})$ in $\mathcal{L}^2(\mathbb{F}, dt \times d\mathbb{P})$, (8.20) is equivalent to requiring

$$\forall \Phi \in \mathcal{A}^2(\mathbb{F}), \quad E\left(\int_0^T \Phi(t)\mathcal{A}F(t, X_t)dt\right) = 0, \tag{8.21}$$

where we can choose $\Phi(0) = 0$ without loss of generality. But (8.21) is well-defined as soon as $\mathcal{A}F(X) \in \mathcal{L}^2(dt \times d\mathbb{P})$. This motivates the following definition.

Definition 8.4.1 (Sobolev space of non-anticipative functionals). *We define* $\mathbb{W}^{1,2}(\mathbb{P})$ *as the space of (i.e., $dt \times d\mathbb{P}$-equivalence classes of) non-anticipative functionals* $F\colon (\Lambda_T, d_\infty) \to \mathbb{R}$ *such that the process* $S = F(X)$ *defined by* $S(t) = F(t, X_t)$ *belongs to* $\mathcal{S}^{1,2}(M)$. *Then,* $S = F(X)$ *has a semimartingale decomposition*

$$S(t) = F(t, X_t) = S(0) + \int_0^t \nabla_M S dM + H(t), \quad \text{with} \quad \frac{dH}{dt} \in \mathcal{L}^2(\mathbb{F}, dt \times d\mathbb{P}).$$

$\mathbb{W}^{1,2}(\mathbb{P})$ *is a Hilbert space, equipped with the norm*

$$\|F\|^2_{1,2} = \|F(X)\|^2_{\mathcal{S}^{1,2}}$$

$$= E^\mathbb{P}\left(|F(0, X_0)|^2 + \int_0^T \text{tr}\big(\nabla_M F(X) \cdot^t \nabla_M F(X) \cdot d[M]\big) + \int_0^T \left|\frac{dH}{dt}\right|^2 dt\right).$$

Denoting by $\mathbb{L}^2(\mathbb{P})$ the space of square integrable non-anticipative functionals, we have that for $F \in \mathbb{W}^{1,2}(\mathbb{P})$, its vertical derivative lies in $\mathbb{L}^2(\mathbb{P})$ and

$$\nabla_\omega \colon \mathbb{W}^{1,2}(\mathbb{P}) \longrightarrow \mathbb{L}^2(\mathbb{P}),$$

$$F \longmapsto \nabla_\omega F$$

is a continuous map.

Using linear combinations of smooth cylindrical functionals, one can show that this definition is equivalent to the following alternative construction.

Definition 8.4.2 ($\mathbb{W}^{1,2}(\mathbb{P})$: alternative definition). $\mathbb{W}^{1,2}(\mathbb{P})$ *is the completion of* $\mathbb{C}_b^{1,2}(\mathcal{W}_T)$ *with respect to the norm*

$$\|F\|_{1,2}^2 = E^{\mathbb{P}}\left(|F(0,X_0)|^2 + \int_0^T \operatorname{tr}\left(\nabla_\omega F(t,X_t)^{\mathrm{t}}\nabla_\omega F(t,X_t)d[M]\right)\right)$$

$$+ E^{\mathbb{P}}\left(\left|\int_0^T \left|\frac{d}{dt}\left(F(t,X_t) - \int_0^t \nabla_\omega F(u,X_u)dM\right)\right|^2 dt\right).$$

In particular, $\mathbb{C}_{\mathrm{loc}}^{1,2}(\mathcal{W}_T) \cap \mathbb{W}^{1,2}(\mathbb{P})$ is dense in $(\mathbb{W}^{1,2}(\mathbb{P}), \|.\|_{1,2})$.

For $F \in \mathbb{W}^{1,2}(\mathbb{P})$, let $S(t) = F(t,X_t)$ and M be the martingale part of X. The process $S(t) - \int_0^t \nabla_M S(u) \cdot dM$ has absolutely continuous paths and one can then define

$$\mathcal{A}F(t,X_t) := \frac{d}{dt}\left(F(t,X_t) - \int_0^t \nabla_\omega F(u,X_u).dM\right) \in \mathcal{L}^2\left(\mathbb{F}, dt \times d\mathbb{P}\right).$$

Remark 8.4.3. Note that, in general, it is not possible to define "$\mathcal{D}F(t,X_t)$" as a real-valued process for $F \in \mathbb{W}^{1,2}(\mathbb{P})$. As noted in Section 7.5, this requires $F \in \mathbb{W}^{2,2}(\mathbb{P})$.

Let U be the process defined by

$$U(t) = F(T,X_T) - F(t,X_t) - \int_t^T \nabla_M F(u,X_u)dM.$$

Then the paths of U lie in the Sobolev space $H^1([0,T],\mathbb{R})$,

$$\frac{dU}{dt} = -\mathcal{A}F(t,X_t), \quad \text{and} \quad U(T) = 0.$$

Thus, we can apply the integration by parts formula for H^1 Sobolev functions [65] pathwise, and rewrite (8.21) as

$$\forall\Phi \in \mathcal{A}^2(\mathbb{F}), \quad \int_0^T dt\, \Phi(t)\frac{d}{dt}\left(F(t,X_t) - \int_0^t \nabla_M F(u,X_u)dM\right)$$

$$= \int_0^T dt\, \phi(t)\left(F(T,X_T) - F(t,X_t) - \int_t^T \nabla_M F(u,X_u)dM\right).$$

We are now ready to formulate the notion of weak solution in $\mathbb{W}^{1,2}(\mathbb{P})$.

Definition 8.4.4 (Weak solution). $F \in \mathbb{W}^{1,2}(\mathbb{P})$ *is said to be a* weak solution *of*

$$\mathcal{D}F(t,\omega) + b(t,\omega)\nabla_\omega F(t,\omega) + \frac{1}{2}\operatorname{tr}\left[\nabla_\omega^2 F(t,\omega)\sigma^{\mathrm{t}}\sigma(t,\omega)\right] = 0$$

on $\mathrm{supp}(X)$ *with terminal condition* $H(\cdot) \in L^2(\mathcal{F}_T, \mathbb{P})$ *if* $F(T, \cdot) = H(\cdot)$ *and, for every* $\phi \in \mathcal{L}^2(\mathbb{F}, dt \times d\mathbb{P})$,

$$E\left[\int_0^T dt\, \phi(t)\left(H(X_T) - F(t, X_t) - \int_t^T \nabla_M F(u, X_u)dM\right)\right] = 0. \qquad (8.22)$$

The preceding discussion shows that any square integrable classical solution $F \in \mathbb{C}^{1,2}_{\mathrm{loc}}(\mathcal{W}_T) \cap \mathcal{L}^2(dt \times d\mathbb{P})$ of the Functional Kolmogorov Equation (8.6) is also a weak solution.

However, the notion of weak solution is much more general, since equation (8.22) only requires the vertical derivative $\nabla_\omega F(t, X_t)$ to exist in a weak sense, i.e., in $\mathcal{L}^2(X)$, and does not require any continuity of the derivative, only square integrability.

Existence and uniqueness of such weak solutions is much simpler to study than for classical solutions.

Theorem 8.4.5 (Existence and uniqueness of weak solutions). *Let*

$$H \in L^2(C^0([0, T], \mathbb{R}^d), \mathbb{P}).$$

There exists a unique weak solution $F \in \mathbb{W}^{1,2}(\mathbb{P})$ *of the functional Kolmogorov equation* (8.6) *with terminal conditional* $F(T, \cdot) = H(\cdot)$.

Proof. Let M be the martingale component of X and $F \colon \mathcal{W}_T \to \mathbb{R}$ be a regular version of the conditional expectation $E(H|\mathcal{F}_t)$:

$$F(t, \omega) = E\big(H \mid \mathcal{F}_t\big)(\omega), \ dt \times d\mathbb{P}\text{-a.e.}$$

Let $Y(t) = F(t, X_t)$. Then $Y \in \mathcal{M}^2(M)$ is a square integrable martingale so, by the martingale representation formula 7.3.4, $Y \in \mathcal{S}^{1,2}(X)$, $F \in \mathbb{W}^{1,2}(\mathbb{P})$, and

$$F(t, X_t) = F(T, X_T) - \int_t^T \nabla_\omega F(u, X_u)dM.$$

Therefore, F satisfies (8.22). This proves existence.

To show uniqueness, we use an 'energy' method. Let F^1, F^2 be weak solutions of (8.6) with terminal condition H. Then, $F = F^1 - F^2 \in \mathbb{W}^{1,2}(\mathbb{P})$ is a weak solution of (8.6) with $F(T, \cdot) = 0$. Let $Y(t) = F(t, X_t)$. Then $Y \in \mathcal{S}^{1,2}(M)$ so, $U(t) = F(T, X_T) - F(t, X_t) - \int_t^T \nabla_M F(u, X_u)dM$ has absolutely continuous paths and

$$\frac{dU}{dt} = -\mathcal{A}F(t, X_t)$$

is well defined. Itô's formula then yields

$$Y(t)^2 = -2\int_t^T Y(u)\nabla_\omega F(u, X_u)dM - 2\int_t^T Y(u)\mathcal{A}F(u, X_u)du + [Y](t) - [Y](T).$$

The first term is a $\mathbb{P}$-local martingale. Let $(\tau_n)_{n\geq 1}$ be an increasing sequence of stopping times such that

$$\int_t^{T\wedge\tau_n} Y(u) \cdot \nabla_\omega F(u, X_u) dM$$

is a martingale. Then, for any $n \geq 1$,

$$E\left(\int_t^{T\wedge\tau_n} Y(u)\nabla_\omega F(u, X_u)dM\right) = 0.$$

Therefore, for any $n \geq 1$,

$$E\big(Y(t\wedge\tau_n)^2\big) = 2E\left(\int_t^{T\wedge\tau_n} Y(u)\mathcal{A}F(u, X_u)du\right) + E\big([Y](t\wedge\tau_n) - [Y](T\wedge\tau_n)\big).$$

Given that F is a weak solution of (8.6),

$$E\left(\int_t^{T\wedge\tau_n} Y(u)\mathcal{A}F(u, X_u)du\right) = E\left(\int_t^{T} Y(u)1_{[0,\tau_n]}\mathcal{A}F(u, X_u)du\right) = 0,$$

because $Y(u)1_{[0,\tau_n]} \in \mathcal{L}^2(\mathbb{F}, dt \times d\mathbb{P})$. Thus,

$$E\big(Y(t\wedge\tau_n)^2\big) = E\big([Y](t\wedge\tau_n) - [Y](T\wedge\tau_n)\big),$$

which is a positive increasing function of t; so,

$$\forall n \geq 1, \quad E\big(Y(t\wedge\tau_n)^2\big) \leq E\big(Y(T\wedge\tau_n)^2\big).$$

Since $Y \in \mathcal{L}^2(\mathbb{F}, dt \times d\mathbb{P})$ we can use a dominated convergence argument to take $n \to \infty$ and conclude that $0 \leq E\left(Y(t)^2\right) \leq E\left(Y(T)^2\right) = 0$. Therefore, $Y = 0$, i.e., $F^1(t,\omega) = F^2(t,\omega)$ outside a $\mathbb{P}$-evanescent set. □

Note that uniqueness holds outside a $\mathbb{P}$-evanescent set: there is no assertion on uniqueness outside the support of $\mathbb{P}$.

The flexibility of the notion of weak solution comes from the following characterization of square integrable martingale functionals as weak solutions, which is an analog of Theorem 8.1.6 but *without* any differentiability requirement.

Proposition 8.4.6 (Characterization of harmonic functionals). *Let $F \in \mathcal{L}^2(dt \times d\mathbb{P})$ be a square integrable non-anticipative functional. Then, $Y(t) = F(t, X_t)$ is a $\mathbb{P}$-martingale if and only if F is a weak solution of the Functional Kolmogorov Equation (8.6).*

Proof. Let F be a weak solution to (8.6). Then, $S = F(X) \in \mathcal{S}^{1,2}(X)$ so S is weakly differentiable with $\nabla_M S \in \mathcal{L}^2(X)$, $N = \int_0^\cdot \nabla_M S \cdot dM \in \mathcal{M}^2(X)$ is a square integrable martingale, and $A = S - N \in \mathcal{A}^2(\mathbb{F})$. Since F verifies (8.22), we have

$$\forall \phi \in \mathcal{L}^2(dt \times d\mathbb{P}), \quad E\left(\int_0^T \phi(t) A(t) dt\right) = 0,$$

so $A(t) = 0$ $dt \times d\mathbb{P}$-a.e.; therefore, S is a martingale.

Conversely, let $F \in \mathcal{L}^2(dt \times d\mathbb{P})$ be a non-anticipative functional such that $S(t) = F(t, X_t)$ is a martingale. Then $S \in \mathcal{M}^2(X)$ so, by Theorem 7.3.4, S is weakly differentiable and $U(t) = F(T, X_T) - F(t, X_t) - \int_t^T \nabla_M S \cdot dM = 0$. Hence, F verifies (8.22) so, F is a weak solution of the functional Kolmogorov equation with terminal condition $F(T, \cdot)$. □

This result characterizes *all* square integrable *harmonic functionals* as weak solutions of the Functional Kolmogorov Equation (8.6) and thus illustrates the relevance of Definition 8.4.4.

Comments and references

Path dependent Kolmogorov equations first appeared in the work of Dupire [21]. Uniqueness and comparison of classical solutions were first obtained in [11, 32]. The path dependent HJB equation and its relation with non-Markovian stochastic control problems was first explored in [32]. The relation between FBSDEs and path dependent PDEs is an active research topic and the fully nonlinear case, corresponding to non-Markovian optimal control and optimal stopping problems, is currently under development, using an extension of the notion of *viscosity solution* to the functional case, introduced by Ekren et al. [22, 23]. A full discussion of this notion is beyond the scope of these lecture notes and we refer to [10] and recent work by Ekren et al. [22, 23], Peng [60, 61] and Cosso [14].

Bibliography

[1] K. Aase, B. Oksendal, and G. Di Nunno, *White noise generalizations of the Clark–Haussmann–Ocone theorem with application to mathematical finance*, Finance and Stochastics **4** (2000), 465–496.

[2] R.F. Bass, *Diffusions and Elliptic Operators*, Springer, 1998.

[3] J.M. Bismut, *A generalized formula of Itô and some other properties of stochastic flows*, Z. Wahrsch. Verw. Gebiete **55** (1981), 331–350.

[4] J.M. Bismut, *Calcul des variations stochastique et processus de sauts*, Z. Wahrsch. Verw. Gebiete **63** (1983), 147–235.

[5] J. Clark, *The representation of functionals of Brownian motion by stochastic integrals*, Ann. Math. Statist. **41** (1970), pp. 1282–1295.

[6] R. Cont, *Functional Itô Calculus. Part I: Pathwise calculus for non-anticipative functionals*, Working Paper, Laboratoire de Probabilités et Modèles Aléatoires, 2011.

[7] R. Cont, *Functional Itô Calculus. Part II: Weak calculus for functionals of martingales*, Working Paper, Laboratoire de Probabilités et Modèles Aléatoires, 2012.

[8] R. Cont and D.A. Fournié, *Change of variable formulas for non-anticipative functionals on path space*, Journal of Functional Analysis **259** (2010), 1043–1072.

[9] R. Cont and D.A. Fournié, *A functional extension of the Ito formula*, Comptes Rendus Mathématique Acad. Sci. Paris Ser. I **348** (2010), 57–61.

[10] R. Cont and D.A. Fournié, *Functional Kolmogorov equations*, Working Paper, 2010.

[11] R. Cont and D.A. Fournié, *Functional Itô Calculus and stochastic integral representation of martingales*, Annals of Probability **41** (2013), 109–133.

[12] R. Cont and Y. Lu, *Weak approximations for martingale representations*, Working Paper, Laboratoire de Probabilités et Modèles Aléatoires, 2014.

[13] R. Cont and C. Riga, *Robustness and pathwise analysis of hedging strategies for path dependent derivatives*, Working Paper, 2014.

[14] A. Cosso, *Viscosity solutions of path dependent PDEs and non-markovian forward-backward stochastic equations*, arXiv:1202.2502, (2013).

[15] G. Da Prato and J. Zabczyk, *Stochastic Equations in Infinite Dimensions*, vol. 44 of Encyclopedia of Mathematics and its Applications, Cambridge University Press, 1992.

[16] G. Da Prato and J. Zabczyk, *Second Order Partial Differential Equations in Hilbert Spaces*, London Mathematical Society Lecture Note Series vol. 293, Cambridge University Press, Cambridge, 2002.

[17] M. Davis, *Martingale methods in stochastic control.* In: Stochastic control theory and stochastic differential systems, Lecture Notes in Control and Information Sci. vol. 16, Springer, Berlin, 1979, pp. 85–117.

[18] M.H. Davis, *Functionals of diffusion processes as stochastic integrals*, Math. Proc. Comb. Phil. Soc. **87** (1980), 157–166.

[19] C. Dellacherie and P.A. Meyer, *Probabilities and Potential*, North-Holland Mathematics Studies vol. 29, North-Holland Publishing Co., Amsterdam, 1978.

[20] R. Dudley, *Sample functions of the Gaussian process*, Annals of Probability **1** (1973), 66–103.

[21] B. Dupire, *Functional Itô Calculus*, Portfolio Research Paper 2009-04, Bloomberg, 2009.

[22] I. Ekren, C. Keller, N. Touzi, and J. Zhang, *On viscosity solutions of path dependent PDEs*, Annals of Probability **42** (2014), 204–236.

[23] I. Ekren, N. Touzi, and J. Zhang, *Viscosity solutions of fully nonlinear parabolic path dependent PDEs: Part I*, arXiv:1210.0006, (2013).

[24] N. El Karoui and L. Mazliak, *Backward Stochastic Differential Equations*, Chapman & Hall / CRC Press, Research Notes in Mathematics Series, vol. 364, 1997.

[25] R.J. Elliott and M. Kohlmann, *A short proof of a martingale representation result*, Statistics & Probability Letters **6** (1988), 327–329.

[26] D. Feyel and A. De la Pradelle, *Espaces de Sobolev gaussiens*, Annales de l'Institut Fourier **39** (1989), 875–908.

[27] P. Fitzsimmons and B. Rajeev, *A new approach to the martingale representation theorem*, Stochastics **81** (2009), 467–476.

[28] F. Flandoli and G. Zanco, *An infinite dimensional approach to path dependent Kolmogorov equations*, Working Paper, arXiv:1312.6165, 2013.

[29] W. Fleming and H. Soner, *Controlled Markov Processes and Viscosity Solutions*, Springer Verlag, 2005.

[30] M. Fliess, *Fonctionnelles causales non linéaires et indéterminées non commutatives*, Bulletin de la Société Mathématique de France **109** (1981), 3–40.

[31] H. Föllmer, *Calcul d'Itô sans probabilités.* In: Séminaire de Probabilités XV, Lecture Notes in Math. vol. 850, Springer, Berlin, 1981, pp. 143–150.

[32] D.A. Fournié, *Functional Itô Calculus and Applications*, PhD thesis, 2010.

[33] E. Fournié, J.M. Lasry, J. Lebuchoux, P.L. Lions and N. Touzi, *Applications of Malliavin calculus to Monte Carlo methods in finance*, Finance and Stochastics **3** (1999), 391–412.

[34] P.K. Friz and N.B. Victoir, *Multidimensional Stochastic Processes as Rough Paths: Theory and Applications*, Cambridge University Press, 2010.

[35] M. Gubinelli, *Controlling rough paths*, Journal of Functional Analysis **216** (2004), 86–140.

[36] U. Haussmann, *Functionals of Itô processes as stochastic integrals*, SIAM J. Control Optimization **16** (1978), 52–269.

[37] U. Haussmann, *On the integral representation of functionals of Itô processes*, Stochastics **3** (1979), 17–27.

[38] T. Hida, H.-H. Kuo, J. Pothoff, and W. Streit, *White Noise: An Infinite Dimensional Calculus*, Mathematics and its Applications, vol. 253, Springer, 1994.

[39] P. Imkeller, A. Réveillac, and A. Richter, *Differentiability of quadratic BSDEs generated by continuous martingales*, Ann. Appl. Probab. **22** (2012), 285–336.

[40] K. Itô, *On a stochastic integral equation*, Proceedings of the Imperial Academy of Tokyo **20** (1944), 519–524.

[41] K. Itô, *On stochastic differential equations*, Proceedings of the Imperial Academy of Tokyo **22** (1946), 32–35.

[42] J. Jacod, S. Méléard, and P. Protter, *Explicit form and robustness of martingale representations*, Ann. Probab. **28** (2000), 1747–1780.

[43] J. Jacod and A.N. Shiryaev, *Limit Theorems for Stochastic Processes*, Springer, Berlin, 2003.

[44] I. Karatzas, D.L. Ocone, and J. Li, *An extension of Clark's formula*, Stochastics Stochastics Rep. **37** (1991), 127–131.

[45] I. Karatzas and S. Shreve, *Brownian Motion and Stochastic Calculus*, Springer, 1987.

[46] A. Kolmogoroff, *Über die analytischen methoden in der wahrscheinlichkeitsrechnung*, Mathematische Annalen **104** (1931), 415–458.

[47] P. Lévy, *Le mouvement brownien plan*, American Journal of Mathematics **62** (1940), 487–550.

[48] X.D. Li and T.J. Lyons, *Smoothness of Itô maps and diffusion processes on path spaces. I*, Ann. Sci. École Norm. Sup. (4) **39** (2006), 649–677.

[49] T.J. Lyons, *Differential equations driven by rough signals*, Rev. Mat. Iberoamericana **14** (1998), 215–310.

[50] J. Ma, H. Yin, and J. Zhang, *Solving forward–backward SDEs explicitly: a four-step scheme*, Probability theory and related fields **98** (1994), 339–359.

[51] J. Ma, H. Yin, and J. Zhang, *On non-markovian forward–backward SDEs and backward stochastic PDEs*, Stochastic Processes and their Applications **122** (2012), 3980–4004.

[52] P. Malliavin, *Stochastic Analysis*, Springer, 1997.

[53] P. Meyer, *Un cours sur les integrales stochastiques.*, in Séminaire de Probabilités X, Lecture Notes in Math. vol. 511, Springer, 1976, pp. 245–400.

[54] A. Millet and M. Sanz-Solé, *A simple proof of the support theorem for diffusion processes.* In: Séminaire de Probabilités XXVIII, Springer, 1994, pp. 36–48.

[55] D. Nualart, *Malliavin Calculus and its Applications*, CBMS Regional Conference Series in Mathematics vol. 110, Washington, DC, 2009.

[56] D.L. Ocone, *Malliavin's calculus and stochastic integral representations of functionals of diffusion processes*, Stochastics **12** (1984), 161–185.

[57] E. Pardoux and S. Peng, *Adapted solutions of backward stochastic equations*, Systems and Control Letters **14** (1990), 55–61.

[58] E. Pardoux and S. Peng, *Backward stochastic differential equations and quaslinear parabolic partial differential equations.* In: Stochastic partial differential equations and their applications, vol. 176 of Lecture Notes in Control and Information Sciences, Springer, 1992, pp. 200–217.

[59] S.G. Peng, *Stochastic Hamilton–Jacobi–Bellman equations*, SIAM J. Control Optim. **30** (1992), 284–304.

[60] S. Peng, *Note on viscosity solution of path dependent PDE and g-martingales*, arXiv 1106.1144, 2011.

[61] S. Peng and F. Wang, *BSDE, path dependent PDE and nonlinear Feynman–Kac formula*, http://arxiv.org/abs/1108.4317, 2011.

[62] P.E. Protter, *Stochastic Integration and Differential Equations*, second edition, Springer-Verlag, Berlin, 2005.

[63] D. Revuz and M. Yor, *Continuous Martingales and Brownian Motion*, Grundlehren der Mathematischen Wissenschaften vol. 293, third ed., Springer-Verlag, Berlin, 1999.

[64] A. Schied, *On a class of generalized Takagi functions with linear pathwise quadratic variation*, J. Math. Anal. Appl. **433** (2016), no. 2, 974–990.

[65] L. Schwartz, *Théorie des Distributions*, Editions Hermann, 1969.

[66] I. Shigekawa, *Derivatives of Wiener functionals and absolute continuity of induced measures*, J. Math. Kyoto Univ. **20** (1980), 263–289.

[67] D.W. Stroock, *The Malliavin calculus, a functional analytic approach*, J. Funct. Anal. **44** (1981), 212–257.

[68] D.W. Stroock and S.R.S. Varadhan, *On the support of diffusion processes with applications to the strong maximum principle*. In: Proceedings of the Sixth Berkeley Symposium on Mathematical Statistics and Probability (1970/1971), Vol. III: Probability theory, Berkeley, Calif., 1972, Univ. California Press, pp. 333–359.

[69] D.W. Stroock and S.R.S. Varadhan, *Multidimensional Diffusion Processes*, Grundlehren der Mathematischen Wissenschaften vol. 233, Springer-Verlag, Berlin, 1979.

[70] H. Sugita, *On the characterization of Sobolev spaces on an abstract Wiener space*, J. Math. Kyoto Univ. **25** (1985), 717–725.

[71] H. Sugita, *Sobolev spaces of Wiener functionals and Malliavin's calculus*, J. Math. Kyoto Univ. **25** (1985), 31–48.

[72] S. Taylor, *Exact asymptotic estimates of Brownian path variation*, Duke Mathematical Journal **39** (1972), 219–241.

[73] S. Watanabe, *Analysis of Wiener functionals (Malliavin calculus) and its applications to heat kernels*, Ann. Probab. **15** (1987), 1–39.

[74] J. Yong and X.Y. Zhou, *Stochastic Controls: Hamiltonian Systems and HJB Equations*, Stochastic Modelling and Applied Probability, vol. 43, Springer, 1999.

[75] J. Zabczyk, *Remarks on stochastic derivation*, Bulletin de l'Académie Polonaise des Sciences **21** (1973), 263–269.

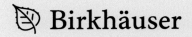 **Birkhäuser** | **birkhauser-science.com**

Advanced Courses in Mathematics – CRM Barcelona (ACM)

Edited by
Enric Ventura, Universitat Politècnica de Catalunya

Since 1995 the Centre de Recerca Matemàtica (CRM) has organised a number of Advanced Courses at the post-doctoral or advanced graduate level on forefront research topics in Barcelona. The books in this series contain revised and expanded versions of the material presented by the authors in their lectures.

■ **Hacking, P. / Laza, R. / Oprea, D.,** Compactifying Moduli Spaces (2015).
ISBN 978-3-0348-0920-7
This book focusses on a large class of objects in moduli theory and provides different perspectives from which compactifications of moduli spaces may be investigated.
Three contributions give an insight on particular aspects of moduli problems. In the first of them, various ways to construct and compactify moduli spaces are presented. In the second, some questions on the boundary of moduli spaces of surfaces are addressed. Finally, the theory of stable quotients is explained, which yields meaningful compactifications of moduli spaces of maps.
Both advanced graduate students and researchers in algebraic geometry will find this book a valuable read.

■ **Llibre, J. / Moeckel, R. / Simó, C.,** Central Configurations, Periodic Orbits, and Hamiltonian Systems (2015).
ISBN 978-3-0348-0932-0
The notes of this book originate from three series of lectures given at the Centre de Recerca Matemàtica (CRM) in Barcelona. The first one is dedicated to the study of periodic solutions of autonomous differential systems in R^n via the Averaging Theory and was delivered by Jaume Llibre. The second one, given by Richard Moeckel, focusses on methods for studying Central Configurations. The last one, by Carles Simó, describes the main mechanisms leading to a fairly global description of the dynamics in conservative systems. The book is directed towards graduate students and researchers interested in dynamical systems, in particular in the conservative case, and aims at facilitating the understanding of dynamics of specific models. The results presented and the tools introduced in this book include a large range of applications.

■ **Alexeev, V.,** Moduli of Weighted Hyperplane Arrangements (2015).
ISBN 978-3-0348-0914-6
This book focuses on a large class of geometric objects in moduli theory and provides explicit computations to investigate their families. Concrete examples are developed that

take advantage of the intricate interplay between Algebraic Geometry and Combinatorics. Compactifications of moduli spaces play a crucial role in Number Theory, String Theory, and Quantum Field Theory – to mention just a few. In particular, the notion of compactification of moduli spaces has been crucial for solving various open problems and long-standing conjectures. Further, the book reports on compactification techniques for moduli spaces in a large class where computations are possible, namely that of weighted stable hyperplane arrangements (shas).

■ **Dai, F. / Xu, Y.,** Analysis on h-Harmonics and Dunkl Transforms (2015).
ISBN 978-3-0348-0886-6
As a unique case in this Advanced Courses book series, the authors have jointly written this introduction to h-harmonics and Dunkl transforms. These notes provide an overview of what has been developed so far. The first chapter gives a brief recount of the basics of ordinary spherical harmonics and the Fourier transform. The Dunkl operators, the intertwining operators between partial derivatives and the Dunkl operators are introduced and discussed in the second chapter. The next three chapters are devoted to analysis on the sphere, and the final two chapters to the Dunkl transform. The authors' focus is on the analysis side of both h-harmonics and Dunkl transforms. The need for background knowledge on reflection groups is kept to a bare minimum.

■ **Citti, G. / Grafakos, L. / Pérez, C. / Sarti, A. / Zhong, X.,** Harmonic and Geometric Analysis (2015).
ISBN 978-3-0348-0407-3
This book presents an expanded version of four series of lectures delivered by the authors at the CRM. The first lecture is an application of harmonic analysis and the Heisenberg group to understanding human vision, while the second and third series of lectures cover some of the main topics on linear and multilinear harmonic analysis. The last serves as a comprehensive introduction to a deep result from De Giorgi, Moser and Nash on the regularity of elliptic partial differential equations in divergence form.

Printed in the United States
By Bookmasters